Make:

Mechanical Engineering
for Makers

Brian Bunnell and Samer Najia

Make: Mechanical Engineering for Makers

By Brian Bunnell and Samer Najia

Printed in Canada.

Published by Make: Community LLC
150 Todd Road, Suite 200, Santa Rosa, CA 95407

Make: books may be purchased for educational, business, or sales promotional use.
Online editions are also available for most titles.
For more information, contact our corporate/institutional
Sales department: 800-998-9938

Publisher: Dale Dougherty
Editor: Patrick Di Justo
Technical Editor: John Manly
Copy Editor: Emily J Gertz
Creative Direction/Design: Juliann Brown
Indexer: Carol Roberts

January 2020: First Edition
Revision History for the First Edition
02-14-2020: Second release
07-31-2020: Third release
04-16-2021: Fourth release

See www.oreilly.com/catalog/errata.csp?isbn=9781680455878 for release details.

O'Reilly Online Learning

For more than 40 years, www.oreilly.com has provided technology and business training, knowledge, and insight to help companies succeed.

Our unique network of experts and innovators share their knowledge and expertise through books, articles, conferences, and our online learning platform. O'Reilly's online learning platform gives you on-demand access to live training courses, in-depth learning paths, interactive coding environments, and a vast collection of text and video from O'Reilly and 200+ other publishers. For more information, please visit www.oreilly.com

How to Contact Us:

Please address comments and questions concerning this book to the publisher:

Make: Community LLC

150 Todd Road, Suite 200, Santa Rosa, CA 95407

You can also send comments and questions to us by email at books@make.co.

Make: Community is a growing, global association of makers who are shaping the future of education and democratizing innovation. Through *Make:* magazine, and 200+ annual Maker Faires, *Make:* books, and more, we share the know-how of makers and promote the practice of making in schools, libraries and homes.

To learn more about *Make:* visit us at make.co.

Table of Contents

Preface

When we sat down to write this book, we wanted to be sure that we didn't create a manual on the use of certain tools or materials, or provide a reference manual for engineering. Rather, we wanted to convey our passion for making that uses techniques many might see as "garage science," but are actually based on real engineering principles. Do you have to be a mechanical engineer to read this book or understand the topics? Is it full of math and formulae? Is it going to be a textbook cure for insomnia? Not at all! Do you like to tinker, to fiddle, to build, to make? Or, maybe you wish to learn how to tinker, fiddle, build, or make? Then this book is for you. This book is intended for everyone who has a passion to make mechanical things (or make them better), whether you are an engineer or not. Engineering school teaches math and theory about engineering, but

rarely provides a hands-on, practical application. That is what this book will provide. It is for all of us backyard- and garage-tinkerers who need just a bit more information to go from a great back-of-a-napkin idea to a glorious finished project.

Within this book, we break down various mechanical engineering concepts to their simplest levels. These concepts are illustrated through hands-on projects, which we strongly encourage you to try on your own. Additionally, we offer *Tracking Further* sections with more in-depth math and theory. These provide opportunities to dig deeper into mechanical engineering theories and principles, while demonstrating interesting and real-world applications of them. Lastly, watch for small sidebars called *Staying on Track*. These are little, practical nuggets of wisdom we've picked up through the years of making, and include specific tips and tricks we have learned through experience, and in some cases trial-and-error.

Both of us, your authors, are degreed mechanical engineers and life-long makers. We have authored articles on makezine.com and have contributed to other publications, such as *Make:* magazine and various websites. We have built airplanes, hovercrafts, pneumatic cannons, RC vehicles, simulators, a variety of tanks, and a plethora of other items. We have torn-apart and repurposed more mechanical devices than we can count! We have had great successes as well as failures in our projects, and will continue to make as long as we are able.

We feel that the innate personality of a practical engineer, tinkerer, or maker is driven by curiosity, and a desire to learn about new and different things. This is exactly what we've kept in mind while writing this book.

We hope you enjoy reading and using this book as much as we enjoyed writing it. — *Brian Bunnell and Samer Najia*

Acknowledgments

BRIAN:
I'd like to start out by thanking my wonderful family for supporting and, at times, putting up with me while writing this book. My wife, Mandy, helped in every aspect and phase of this process. She was my transcriber, wordsmith, sounding board, sanity check, and devil's

advocate: challenging my ideas, questioning my technical explanations, listening to my scattered thoughts, and sometimes re-writing entire sections for clarity. Thank you!

My kids, Kenna and Evan, not only gave me the time to write this book, but were also actively involved in the process. Kenna, my daughter, drew several of the illustrations, and Evan, my son, served as the "chief tester" of nearly all of the projects. Great job, kiddos!

I'd like to also thank my dad, Richard Bunnell, for being my ultimate mentor in building, tinkering, making, and life. Second only to my dad, John Rob Holland was the most influential person of my childhood in a maker sense. Thank you, John Rob, for fostering the Young Inventor in me. Also, I want to sincerely thank Tom Heck for introducing me to the Maker Movement and involving me in the local maker community.

I want to express my gratitude to my mentor, Dr. Jay Huebner, for allowing me to work in his lab at the University of North Florida. Thanks also to Dr. Wei Chen, for allowing me to work with her in her lab and ultimately helping me earn my way through engineering school, all while doing interesting work for her and various other professors at Clemson University. I truly appreciate that you both have believed in me, and provided me with an optimal environment to create, learn, and explore.

I want to sincerely express my thanks to my good friend, fellow mechanical engineer and maker, John Manly. He reviewed the book from both technical and flow standpoints, and graciously contributed to the book by authoring several sidebars.

Finally, thanks to Samer Najia for all of his contributions in this book — but I appreciate your friendship even more.

SAMER:
If it weren't for the support of my fabulous family, both immediate and extended, I would still have built all my projects, but no one will ever know about them. Thanks to them, I was able to disappear into the garage for days on end. My family inspires me to create and they are my first "beta testers". Between my wife Sanja taking some of the pictures for this book, and my kids Hanna and Jordan showing me what non-engineers are expecting, I continue to learn how to make things, and make them better.

My nephew, Max Thorson, was instrumental in getting my tank moving (literally). If it weren't for him, I'd still be guessing about a lot of things.

I'd also like to thank my many friends at NOVA Labs (nova-labs.org) and especially Ted Markson. I bounced a lot of ideas off Ted and, thanks to him, learned a lot about what real things should be like. I have to single out Brian Jacoby, who encouraged me while I was building the tank and flight simulator, and gave me the opportunity to exhibit my projects at the NOVA Mini Maker Faire.

Mr. Tim Van Milligan of Apogee Rockets deserves a tip of the hat here as well. Tim helped me focus some of my rocketry projects, and featured ome of those designs on his website apogeerockets.com.

I've learned a lot from Griffin Paquette. As of this writing, Griffin is a mechanical engineering college student, but really he might as well be a Professor, PhD and all. Thank you, Griff, for your patience and guidance with all things related to 3D printing.

Jason Sanchez is a brilliant electrical engineer and "drone-iac." He helped me work out the electrical switch controls for the Adult-Size Tank without frying my fingers or electrocuting myself. Without his help, my tank would be a static sculpture. Thank you, Jason. I still think you should build much bigger things.

Needless to say, Brian, my co-author and Quality Assurance System, has made so much possible. Had we not somehow connected, this book might never be what you see today. Brian is my friend and my maker "twin." It's probably better for both of us to live several hundred miles apart. If we were next door neighbors, Cape Canaveral would probably be rivaled by what we put together in our backyards. Thank you, Brian! You've been an inspiration and an endless source of ideas.

Brian's Story

All the neighbor kids have gathered around. It's time for the first test of the latest project from the Bunnell workshop. My dad and I (I'm 12 years old) have constructed an electric vehicle made of a diverse, somewhat disparate conglomeration of materials. The frame is made of aluminum channel originally intended for use building screened porch enclosures. The wheels are from bicycles. The motor is a starter motor out of an old Toyota pickup truck. And the transmission is from a junked riding lawnmower.

The transmission will serve as the point at which we will pick back up on the story...

I'm sitting there, ready to go, clutch depressed and motor running. I turn my head and look over my shoulder to make sure that I am clear to proceed in reverse. I pop the clutch in one swift fluid motion, and BAM! I slam straight into the garage door directly in front of me! In our excitement and (possibly) haste, my dad and I have installed the transmission backward. Our electric vehicle has three speeds in reverse, and one in forward. Oops!

One might consider that first test run an utter failure. But considering the vehicle did at least move, and the test was immensely entertaining for those gathered, it could also be considered a success.

After somewhat fixing the dented and scuffed garage door, my dad and I proceeded to reverse the gearing, thereby giving the vehicle its intended three speeds forward and one in reverse. The second test run went much more smoothly, though a lot less memorably.

The moral of this story is that from the initial mistake of installing the transmission backward, I learned several important things. I learned how to reverse the input to a transmission, and how to repair a dented and scuffed garage door. Most importantly for making, though, I learned how to laugh at my mistake, figure out how to fix it, and implement that fix. Throughout this book, I will be sharing with you other stories along the same lines as this one, because some of the best lessons in making I have learned, and you will learn, are derived from "mistakes." Those "oops" moments seem to have a way of burning into our memories, and provide for interesting, sometimes funny stories to share with others. We also tend to remember and take pride in how we recover from those snafus. To me, making is about the journey, about the successes and failures, but most importantly, about sharing with others. So, if you will, allow me to take you on a journey in making and share with you both my successes and failures, and what I have learned along the way.

Samer's Story

I recall that when I was around 12 or 13 years old, I decided that I needed to build an all-terrain vehicle or ATV. Back then I lived in Europe, and stuff like that was not something any kid might be able to get on hand. Those were the days of motocross bicycles, as well as the thin-wheeled racers most kids rode around on. I wanted more, much more. Tinkering for me started early, and armored vehicles were a particular fascination: tracked monsters, six-wheeled beasts, anything that could go anywhere (besides airplanes) caught my eye. My first foray into "engineering" as a maker was to make my ATV a reality by cobbling together bits of toys discarded by the neighbors: pram (or strollers as we know them here) wheels, miscellaneous nuts and bolts, wood from old chairs or tables. All led to a sort of sled that I lay down on and steered much like a toboggan. Danger and injury are not things kids think of at that age, and I was no different. It wasn't long before my "kart" grew walls and a chair. Then I managed to destroy it, in an event I declined to tell my parents about.

Some dreams never die, and decades later (after earning a degree in mechanical engineering and building an airplane), I had some leftover scrap metal from the airplane. So I used all that aircraft-grade aluminum to make a Go-Kart. Zipping along my neighborhood at 30mph, at just 4" above the ground, was not my brightest idea. Not to mention that I did not take into account under-engineering my rear shaft, which at speed flexed enough to throw the chain drive on the right rear wheel out of alignment, causing an abrupt (and scary) loss of power.

I had rebuilt that Go-Kart twice by the day I met Brian, after seeing his Tank design. The Go-Kart went from half-track to full tank — and yes, my inner 13-year-old felt a sense of accomplishment at perfecting one long-lasting project.

Dreams should never be surrendered to "reality." You just have to have the determination to "engineer" a compromise, even if it takes you forever. Never be afraid to fail, but also never give up. Any dream worth its salt is worth pursuing to its end. So now, let's do that...

Foreword

Why do we build things?

In "The Cathedral and the Bazaar," Eric Raymond wrote that open-source developers write code and share it openly because they are "scratching their own itch." They have something in mind that needs to be done. So they just do it, and then share it with others. Some people are skeptical of this idea of "scratching your own itch." It makes some business people scratch their heads, asking: "Why do it if you're not being paid?" They don't understand that there are people of all ages who just build things because that's what they love to do.

Scratching your own itch was also behind the Maker Movement, which started when lots of people independently began sharing the projects they built based on their own interests and abilities. Makers have many reasons why they build something, but the more interesting question is how they do it. We like to learn from each other how to build things.

Ever since I first used the word "maker" intentionally in *Make:* magazine, I have been fascinated by the variety of projects that makers have worked on in backyards, in garages, and at kitchen tables. These real-world projects use cheap hardware and sensors, often embedding them in physical things to create robots, self-driving cars, wearables, and interactive art. Each year I can discover new and surprising projects at Maker Faires around the world.

In the magazine, makers have shared the how-to procedures for their projects, so that others might learn from them and build that project, or others like it. Makers have talked about the tools they use and the different parts that they source. As computer-assisted design (CAD) have become easier to use, and new machines such as 3D printers have made it possible for individuals to do manufacturing in small lots. Makers were the first to realize that a prototyping revolution was taking place, where going from idea to prototype was faster and cheaper and more accessible than ever.

If there is an essential quality for a maker, it is curiosity. Makers are interested in figuring out how things work. They are good learners, largely because they are motivated to do something with what they learn. The best makers expand their practical skills but also deepen their conceptual knowledge. However, as projects become more complex and ambitious, the learning curve can be more challenging. Some people fall off. Others do it as a profession.

This book, *Mechanical Engineering for Makers*, is intended to help makers continue continue their journey to do more by exploring the basics of mechanical engineering. Makers can benefit from understanding how mechanical engineers think about machines and systems. Nonetheless, the goal is to apply this thinking in a typical maker project. The authors, Brian and Samer, are makers who understand why people work on projects in a garage for fun. They are also mechanical engineers who have knowledge and experience gained from the university and the workplace.

By intention, this book is not a textbook. Nor is it a replacement for a course of study in mechanical engineering. It's really a crossover, a new bridge between making and mechanical engineering, between amateur and professional. You can go back and forth as many times as you want, and bring things of value with you. It won't make you an expert. But it will generate insights and reward your curiosity. You will learn new ways to approach problems. Ideally, this book may introduce young makers to the prospects of an engineering discipline and a fulfilling career. After all, these authors were once kids who made things.

With detailed procedures, images, and diagrams, as well as a little math, the book is accessible to non-technical readers interested in learning more about how to build things. Samer and Brian explain the theory in clear language and show the practice through a wide range of projects. I hope it helps you have more fun and be more satisfied with scratching your own itch, whether at work or play. May you tackle more interesting projects!

— *Dale Dougherty*
Founder of *Make:* magazine and Maker Faire

1

Mechanical Engineering – A Maker's Perspective

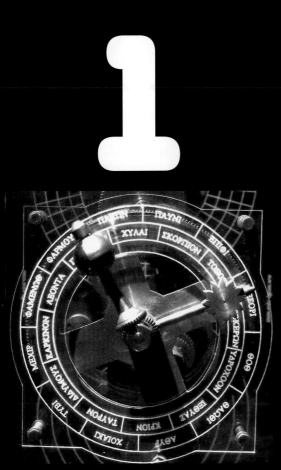

A reconstruction of the Antikythera Mechanism – beauty in engineering. Image was taken by Marsupium Photography, Berlin 2012

Engine illustrating many mechanical subsystems and materials

Mechanical engineering is the branch of engineering concerned with the design and analysis of mechanical systems. It is one of the broadest of the engineering disciplines. Mechanical engineers must often consider physics, mathematics, mechanisms, materials, thermodynamics, and power systems such as hydraulics and pneumatics in their designs. Elements of other engineering disciplines are also often found in mechanical engineering projects. For instance, it is common for mechanical engineers to incorporate aspects of electrical and electronics engineering in their designs – including power sources, control systems, signal processing, and computing. Mechanical engineering may also cross over into other engineering disciplines such as civil, aeronautical, biomechanical, chemical, and electrical engineering.

Becoming a mechanical engineer requires four to eight years of university-level study in topics including calculus, physics, drafting, fluid mechanics, and much more. As a maker or hobbyist, you probably aren't interested in delving too deeply into these topics. You just want to know enough to get better at developing the projects you've dreamed up for yourself. That's what we're here to help you with!

Basic Topics of Mechanical Engineering

Let's take a brief look at some of the main topics in mechanical engineering, which you'll be putting into practice as you work your way through this book. We'll begin with simple machines, since they are the building blocks of mechanical systems and — as you'll see — we will return to them throughout the book. A **simple machine** is defined as a device that changes the direction and/or magnitude (amount) of a force. When considering the change in force a simple machine generates, mechanical engineers talk in terms of **mechanical advantage** (see *Tracking Further: Mechanical Advantage* on page 2).

It's quite easy to calculate the mechanical advantage that a machine generates! It is the ratio of the *output* force to the *input* force. For example; say a machine requires you to push a lever with a force of 20lbs to lift a load of 100lbs. Thus the output of the system is 100lbs, while the input is only 20lbs. Dividing the output force (load) by the input force (effort) we get the number 5. So, for this system, the mechanical advantage is 5, meaning that for every pound of input force (or effort) you get 5 pounds of output force.

In most real-world engineering calculations, engineers must make certain assumptions in order to solve a given problem. Can you deduce the major assumption that we made in this example of mechanical advantage? The assumption was that the system was perfectly efficient, that it had zero friction throughout. We assumed that the mechanism did not consume any energy internally while it was being operated. In engineering terms, we call this perfectly efficient system an **ideal system**.

If the mechanical advantage is calculated to be greater than 1, we say that there is an **amplification of force**. If the calculated mechanical advantage is exactly 1, the input to the system is exactly equal to the output, therefore there is *no* mechanical advantage. If you calculate a mechanical advantage of less than 1, then the system actually *reduces* the amount of output force relative to the input effort.

Why would you ever want to design a system that has a mechanical advantage of less than 1, which requires more force to be applied to it than the force you get out of it? The answer is that what you lose in output force, you gain in greater speed and/or displacement. Look at bicycle chains and sprockets. If the pedals are turning a big sprocket and the chain leads to a smaller sprocket at the rear wheel, your mechanical advantage is less than 1, and on a straight road you would go faster, (although getting started would have you huffing and puffing a lot more). Now, what if you shifted gears so that you have a smaller sprocket at the pedals and a bigger sprocket at the wheel? Your mechanical advantage is closer to 1; now, you can start the bicycle moving from a standstill much easier. Typically, a bicycle will not have a sprocket at the crank (pedal) that is smaller than any of the sprockets at the wheel. That is why this example can never have a mechanical advantage greater than 1.

We'll discuss mechanical advantage more in the chapters on levers, pulleys, and gears.

Using a crowbar; amplification of force

Using a bypass lopper; amplification of force

Bicycle chain and sprocket: reduction of force

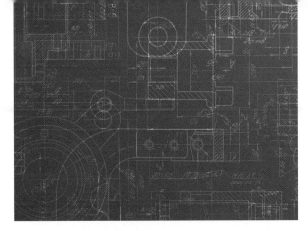

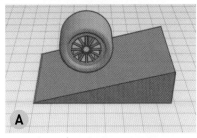

A

3D model of an inclined plane

B

Inclined plane in the form of a ramp

Simple Machines

There are six types of **simple machines**. You probably learned about them in your high school physics class, but we'll review them here. They are the inclined plane, the wedge, the wheel-and-axle, the lever, the pulley, and the screw.

An **inclined plane** (Figures A and B) is a rigid surface that is oriented at an angle with one side higher than the other. A ramp is a common example of an inclined plane, and it is classified as a simple machine because it reduces the amount of force required to move a load a given vertical distance. (See *Tracking Further: Conservation of Energy and Work* on page 5 for an engineering example that uses the inclined plane.)

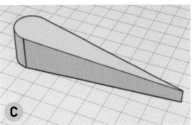

C

3D model of a wedge

A **wedge** (Figures C and D) is nothing more than an inclined plane that can be moved. A door stopper is an example of a wedge. A wedge's purpose is to exert a force between two objects through its surfaces to either force the objects apart or to secure them. Contrast this with an inclined plane, where force is typically exerted by an object that is located *on* the inclined plane.

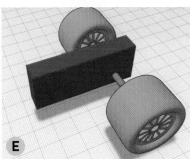

D

A wedge in the form of a door stop

A **wheel-and-axle** (Figures E and F) is a system made up of a disk-shaped object that can rotate about a rod. A doorknob is an example of a wheel-and-axle. A rotational force exerted by your hand on the outside of the doorknob imparts a rotational force on the doorknob's central axle. This central axle then conveys that force to a mechanism that retracts the door latch. The force the axle exerts on the latch mechanism is greater than the force you must exert to rotate the knob, because the size of the knob is much larger than the diameter of the axle. This gives you a positive **mechanical advantage** or force amplification. This force amplification is an example of what makes a wheel and axle a simple machine.

E

3D model of a wheel-and-axle

F

A wheel-and-axle in the form of a doorknob

3D model
of a lever

A Lever in the
form of a toy
seesaw

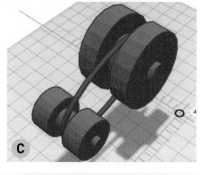

3D model
of a pulley

Traditional cable/
rope and pulley
system

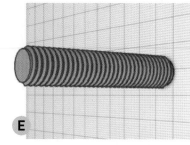

3D model
of a screw

A screw in the
form of a
turnbuckle

A **lever** (Figures **A** and **B**) is a device made up of a rigid body with three basic elements: a load, a pivot point (or **fulcrum**), and an applied effort. Levers can change the direction of force and can change the effort force relative to the load force. A fulcrum is nothing more than the pivot point around which a lever rotates. Think of a seesaw. The fulcrum is the hinge or pivot in the middle of the see-saw.

A **pulley** (Figures **C** and **D**) is a wheel and axle system around which a belt, chain, rope, or cable can move. Sometimes the belt or chain is in a groove in the wheels, and sometimes the wheels have 'lips' that prevent the chain or belt from sliding off. By wrapping a rope over a single pulley, you can change the direction of a given pull force. If multiple pulleys are used together, you can also change the system's output force relative to the input force.

A **screw** (Figures **E** and **F**) is a wedge that has been twisted into a helix wrapped around a cylindrical body. By rotating a screw in matching screw threads, a very large axial force can be generated. An axial force is a force in the direction of the screw. This means any force applied is in the direction the screw-tip is pointing.

Throughout this book, we will be discussing several of these simple machines in significant detail. We will look at both how to design systems with these machine elements, and also explore how to combine these simple elements into more complex configurations. Additionally, in some cases, we will discuss how to analyze these in a mathematical sense.

Materials

In designing components and mechanical devices, mechanical engineers must consider which materials to use. They review properties such as weight, durability, flexibility to determine the best material for a specific purpose. The material you use to make your project will need to be able to support the load (weight) and stresses (tensile, compression, shear, torsional and bending) that will be applied to it.

A mechanical engineer also considers how well the material can be formed into the component that is required. This refers to the idea of **manufacturability**. Some materials, such as metals, form into a variety of geometries and shapes but require forging or special tools (cutters, lathes, mills) to get the desired geometry. Other

Tracking Further: Conservation of Energy and Work

is the concept behind many calculations in mechanical engineering (and in fact, just about every scientific discipline). It simply states that the total energy in an isolated system is neither created nor destroyed; it simply changes form. So what does conservation of energy have to do with, for example, the simple inclined plane? If the total energy (in the form of mechanical energy or work) is constant, how does the inclined plane reduce the amount of force required to lift a load? The answer lies in the equation of work for this particular system. **Work** is equal to the force exerted, multiplied by the distance traversed.

For example, we need to lift a 50lb box of parts up onto a platform 4ft off the ground, as illustrated in Figure **G**. The work required is 4'*50lbs or 200'*lbs. Now, let's suppose that there is a ramp next to the platform at a 14.5° angle, up which we can carry our 50lb load to get to the 4' high platform, as illustrated in Figure **H**. The distance we need to walk up the ramp is the hypotenuse of the triangle described by our ramp angled at 14.5° and elevated 4' at one end. That works out to 16'. Since we know the amount of conserved energy or work, we can determine the force required by dividing the work by the distance traveled: 200'*lbs/16'=12.5lbs. That gives us a mechanical advantage of 50 lbs/12.5lbs = 4. It takes only one-fourth of the force to lift the box of parts using the ramp that it would take without it.

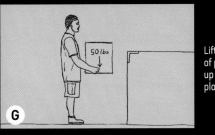

Lifting 50lb box of parts straight up onto a 4' platform

G

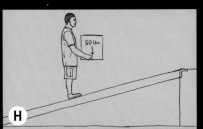

Lifting 50lb box of parts onto a 4' platform using a ramp

H

materials, such as plastics, can also be manipulated into various shapes, yet may be easier to work with because standard tools (drills, band/scroll saws, and bending/shaping tools) are enough to produce the required outcome. As a maker, this aspect of materials selection is important.

Another materials consideration is how the material can be fastened to itself or other components. Materials such as wood are easily fastened to themselves and nearly any other material via mechanical means like screws, nails, and bolts. Steel can also be fastened mechanically via bolts, but steel sometimes requires welding for certain joints. Plastics can be screwed and bolted together but can also be glued or chemically welded.

Other properties that sometimes need consideration are weight, wear properties (how how quickly the the material will break), **lubricity** (how slippery a material is, especially for applications where smooth rotation is required), and the material's ability to withstand the environment (such as exposure to moisture, temperature, chemicals, radiation, wind, food, kids, etc.).

Finally, an engineer must also consider cost (and availability) of materials. Sometimes, there are a few material options that will work to create the component, but one is much more expensive than the other. Other times, there is a material that is best for the job and another that is okay for the job, but the okay material is sitting in your garage taking up space. For us makers, often the "just okay but already available" status trumps other considerations!

A simple seat made from a tree stump and a plank

A forged metal hook

An aluminum drive hub machined on a lathe

Server rack rails turned into a support beam

Before we leave this topic, let's step back for a moment and take a closer look at stress, which we mentioned at the beginning of this section. Understanding the kinds of stress your project may be subjected to can be a critical factor in your choice of materials.

Stress

Stress, in the mechanical sense, is a measure of the internal forces an object is subjected to, divided by the area on which the forces act. In the English system, stress is measured in pound-force (lbf) per square inch, or psi. This is the pressure that results from one pound-force applied to an area of one square inch. The air pressure in your tires is usually measured in psi. In the International System of Units (SI), the unit of measure for stress is the pascal (Pa), measured in newtons per meter squared (N/m^2).

Tracking Further: What is Newton?

A newton (N) is the SI unit of force. Because of his work in classical mechanics, this unit of force was named after Sir Isaac Newton. A newton is defined as the force required to accelerate a mass of 1 kilogram by 1 meter per second squared (1N=kg*(m/s^2)).

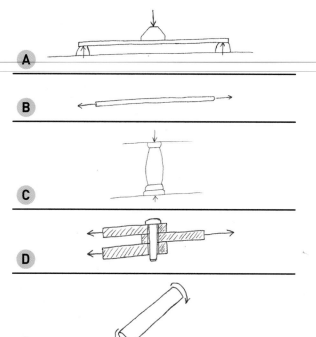

There are various types of stress. Here are a few that you are likely to encounter.

- **Bending Stress** (Figure **A**) is the stress that occurs when a force is applied to a long, slender part perpendicular to the longitudinal axis of the part. Think of a horizontal beam with a vertical load applied to its center. The beam will try to bend due to this load.

- **Tensile Stress** (Figure **B**) occurs when an object is being pulled in two opposite directions. Think of what a rope "feels" during a game of tug-of-war.

- **Compression Stress** (Figure **C**) is the type of stress an object is subjected to when the forces are applied toward each other from opposite directions. These forces are trying to push the object together. Think of a column supporting the weight of a roof.

- **Shear Stress** (Figure **D**) can be thought of in terms of the stress an object experiences when two sides of the object are being slid in opposite directions relative to each other. An example of something in shear stress is a bolt holding two plates together while the plates are being pulled in opposite directions. The bolt experiences shear stress.

- **Torsion Stress** (Figure **E**) is the stress develops when something is being twisted. A driveshaft in a car predominately experiences torsional stress while the car is being driven.

- **Combined Stress** (Figure **F**) is a combination of more than one of the five stress types listed above. In most real-world situations, components will experience combined stress. For example, a stop sign post will experience significant bending and torsional stresses during a windstorm.

A storm-damaged stop sign illustrating combined stress

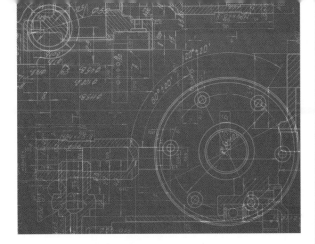

Static Structures and Dynamic Systems

Static structures are those that have no motion relative to a reference point. For example, a child's fort (Figure **G**), as well as a less obvious example, the frame of an RC car are both static structures. Static structures are designed to have no moving parts: the structure's geometry is intended to remain motionless, be stiff, and not deform under the loads that it will see during operation. For example, as the RC car moves across terrain like rocks and sticks, the frame will remain a stable, solid platform on which all other components (motor, servos, battery, etc.) are anchored.

Dynamic structures (or systems) differ from static structures in that they are designed to accommodate relative movement. In our RC car example, the suspension linkage carrying the rotating axles and wheels is a dynamic system (Figure **H**). A lot of maker projects incorporate dynamic systems involving movement between the pieces. These moving systems are, however, always tied into a static structure. Every dynamic system has a static structure in the form of a frame, base, foundation, etc.

Although it is not a common problem, it is very important to mention one area of concern that must be considered in both static and dynamic structures. **Harmonics** (amplitude and frequency) of motion can cause **vibration** in a system. Every mechanical system has a natural frequency based on the mass and stiffness of the system. If a mechanical system is vibrated at its natural frequency, the system will begin to oscillate (move back and forth). This oscillation will build until the system fails or breaks apart (Figure **I**). Check out the video of the collapse of the Tacoma Narrows Bridge online (wikipedia.org/wiki/Tacoma_Narrows_Bridge) to learn just how destructive harmonics can be!

G A fort illustrating a static structure

H RC car suspension illustrating a dynamic structure

I Harmonic oscillation destroyed the Tacoma Narrows Bridge in 1940

A Shock absorber illustrating a thermal/fluid system

Thermal/Fluid Systems

Thermodynamics is the study of energy transfer via heat.

Fluid mechanics is the study of how fluids flow, as well as of the systems that use fluid power. Fluids can be liquids or gases. Liquids are treated as incompressible fluids, meaning that they can't change volume, while gases are considered compressible fluids that can change volume within a certain fixed geometric (i.e., non-flexible) container.

Thermal/fluid systems is a generic term for systems that transfer energy via fluids. The subject of heat transfer combines aspects of both thermodynamics and fluid mechanics.

A typical illustration of thermal/fluid systems is in the application of hydraulics, which is commonly found in a vehicle's shock absorber (Figure **A**). The shock absorber contains a fluid (in this case a liquid, oil, which is incompressible) that absorbs the force of compression (such as when a wheel hits a bump), passing the energy to the fluid by forcing it with a piston through small passages and valves. The resistance caused by forcing the fluid through the passages causes energy to build up in the form of heat. Once the vehicle's wheel gets past the bump, the fluid can move back out of the passages and push the wheel back into its "pre-bump" position. The thicker (more viscous) the fluid, the larger the effort required to

force it through the passages. This effort defines a shock absorber's ability to smooth the ride.

An example of thermal/fluid systems using compressible fluids can be illustrated with a pneumatic cannon (i.e., a potato gun), as shown in (Figure **B**). Compressed air from a pressurized chamber is quickly exhausted into the cannon's barrel, which is plugged closed with an object (usually a potato). The sudden force exerted by the compressed air pressurizes the barrel and forces the object out as a projectile. In the process, mechanical energy is transferred to the object, but there is also heat transfer taking place. The rapid decompression (or expansion) of the air causes a dramatic decrease in temperature. After being shot, a cannon often looks like it is smoking. This is due to moist air from the environment hitting the dew point, causing water droplets to condense and create fog (Figures **C** and **D**).

Another way to illustrate the concepts of thermal/fluid systems is through the example of an electric motor. An electric motor generates heat, which needs to be dissipated or removed from the system somehow. Usually, this is done via a heat sink, which often consists of a block of aluminum with fins machined into it (Figure **E**). Aluminum has a good coefficient of thermal conductivity (or, in plain English, it conducts heat very well). The aluminum in the heat sink, therefore, draws heat *away* from the motor. The fins, in turn, increase the surface area in contact with the air around it, increasing the transfer of heat from the motor into the surrounding air.

Heat can be transferred via conductive heat transfer, convective heat transfer, or radiant heat transfer. The heat energy transferred between the motor and the aluminum is **conductive** heat transfer; that is, the transfer of heat due to contact of two or more solid bodies at different temperatures. The heat transferred between the fins and the air is **convective** heat transfer (contact of a fluid body and a solid or another fluid body at different temperatures). In some cases where more heat needs to be dissipated, a fan is used to force air over the motor and the heat sink. This is known in engineering terms as forced convection.

B   Pneumatic cannon showing projectile in barrel

C Pneumatic cannon "smoking"

D Pneumatic cannon "smoking"

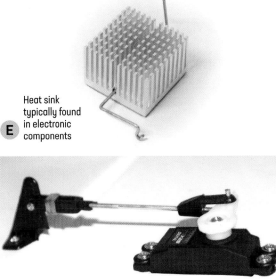

Heat sink typically found in electronic components

E

G Home doorbell mechanism showing a solenoid

F Hobby servo in RC plane

Unlike conductive and convective heat transfer, *radiant* heat transfer does not need a medium (solid or gaseous object) to transfer thermal energy from one object to another. The thermal energy is transferred via the electromagnetic waves that all objects emit. Some of the heat built up in the motor as it runs will be emitted as electromagnetic waves, which will then be absorbed by surrounding objects such as enclosures or other physical components close to the motor. This is how radiant heat transfer helps draw heat away from the motor.

For makers, having a general understanding of energy transfer via fluids will help you create designs to take advantage of these concepts, and also help you further understand how other mechanical systems work.

Electrical Engineering and Electromechanical Systems

Mechanical engineers must learn the basics of electricity and electrical systems. In fact, mechanical engineering students are required to take a general electrical engineering course in school. Basic knowledge of electrical systems is required for mechanical engineers to design and work with electromechanical systems that give power and actuation to projects. Makers also routinely work with electromechanical systems such as motors, servos (Figure **F**), and solenoids (Figure **G**). Therefore, when considering mechanical projects, a maker will most likely need some basic knowledge of electrical components and systems.

We will be discussing more electrical-engineering-related concepts and components in Chapter 9.

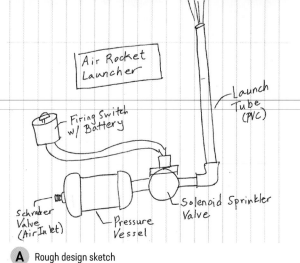

A Rough design sketch

Design and Design Tools

Design is the process an engineer goes through to figure out how something should be made. When designing a new device, mechanical engineers must consider how its components will be made and how they will be assembled into a larger system.

In designing more complicated mechanical systems, it is often necessary to break them down into simpler elements. As we discussed earlier, simple machines are the elements mechanical engineers use as fundamental building blocks in more complicated systems.

DRAWING TOOLS

A design tool combo that we find absolutely indispensable in both our professional and our hobbyist/making endeavors is pencil and paper. Even in our daily, professional practice as mechanical engineers, we start nearly all concepts with simple, back-of-the-napkin sketches (Figure **A**). Sketching helps flesh out ideas by showing the relative scale and placement of parts. It also provides a tangible illustration that can be shown to others for feedback and suggestions.

Moving beyond pencil and paper, you may find that **computer-aided design** (CAD), c**omputer-aided manufacturing** (CAM), and **finite element analysis** (FEA) are useful design and manufacturing tools.

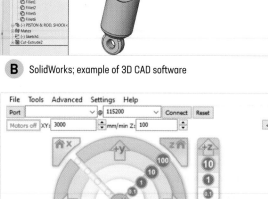

B SolidWorks; example of 3D CAD software

- **Computer-Aided Design (CAD)** software is used to model mechanical systems on a computer. Examples of professional CAD software include AutoCAD for 2D drawing and SolidWorks (Figure **B**), Inventor, and ProE for 3D modeling. As a maker, you will find that there are many free-to-low-cost software tools available, including TinkerCAD (tinkercad.com), which is entirely in the cloud, FreeCAD (freecadweb.org), OpenSCAD (openscad.org), ExpertCAD (amt-software.com/ExpertCAD.html for 2D], and SketchUp (sketchup.com). Learning to use any of these programs will take your drawings to the next level!

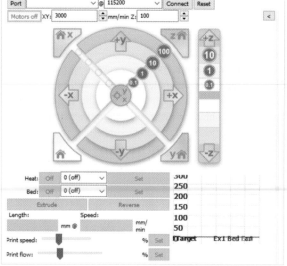

C Printrun/Pronterface; example of 3D printing controller software

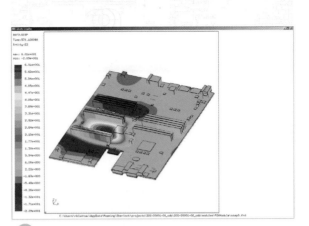

- **Computer-Aided Manufacturing (CAM)** software is used to manufacture the components you designed in your CAD program. Examples of professional CAM tools are Computer Numeric Controlled (CNC) machining centers such as lathes and mills, laser cutters, water jet cutters, and plasma cutters. For home use, makers can use 3D printers as CAM tools (Figure **C**). *Make:* magazine has featured many 3D printers and publishes a 3D printer roundup every year (see, for instance, makezine.com/comparison/3dprinters).

D Sherlock Automated Design Analysis™; example of FEA

- **Finite Element Analysis (FEA)** is typically used only in professional settings to analyze stress/strain on critical components (Figure **D**). Most makers will not need to perform these analyses. However, if you are interested in learning more about FEA, see simscale.com/docs/content/simwiki/fea/whatisfea.html.

You may be surprised to learn that professional mechanical engineers and makers alike commonly use spreadsheet apps, such as Excel and Google Sheets, for conceptualizing design and manufacturing. Spreadsheets are especially helpful for managing sets of data and information, such as numbers, material lists, and calculations, etc.

Manufacturing

Connected to the concept of design is that of manufacturing. This term usually refers to large-scale production, but here we'll also use it to refer to the simple act of fabrication. As a maker, you must often temper your design concepts with the practical considerations of manufacturing; that is, what tools and materials are available to you, the costs, and so forth. Throughout this book, we'll examine the processes of design and manufacturing with sample projects that we encourage you to make yourself. You'll see how we adapt our ideas to available materials, and you'll get the opportunity to do the same.

A mechanical engineer must consider more than just the design, form, and concept of a system, but also *how* to make it. For instance, if a laser cutter is available, the engineer may choose to design components out of sheet metal that can be cut out using a laser and bent into a specific form, rather than designing the part to be machined out of a solid block of metal. As makers, we should ask ourselves questions such as, "Do I have the tools available to make these parts? Do I have the skill or expertise needed to use those tools? What tolerances

Samer's small workbench where detail work and 3D printers come to life

Sometimes the project IS the workbench – Samer's ATV in his garage

Alan Baum

must I stay within, and are those restrictions feasible? Do I have (or can I readily get) the required materials?" You may need to modify your design based on the answers to these questions; a give-and-take based on feasibility and limitations.

Cost will always be a factor in your designs. If a design or the materials needed are outside of your budget, you'll need to explore whether alternatives exist that may provide the same function for less money. Can the design be altered to cost less while remaining safe and functional? The make-versus-buy equation may come into consideration: Will the design be more cost-effective, or time-effective, if you purchase part of a sub-system instead of making it from scratch? Managing costs against function and feasibility is as important (and maybe even *more* important) to makers as to professional engineers. If there is a way to keep costs down while maintaining the function and safety, then that is likely the best approach.

Makers, like mechanical engineers, will work through multiple design and manufacturing scenarios, analyzing each one for functionality, ease of manufacturing, and cost, before finalizing design decisions.

Laser cut sheet metal frame of Brian's Drift Trike Industrial that he built from Alan Baum's plans

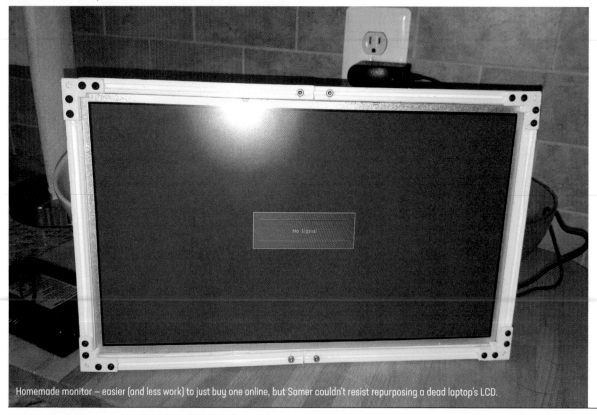

Homemade monitor – easier (and less work) to just buy one online, but Samer couldn't resist repurposing a dead laptop's LCD.

Two Tales of Maker Inspiration

We'll close out this chapter with two tales of maker inspiration that led to the partnership which has resulted in the book you hold in your hands.

THE TOT-SIZE TANK IS BORN

When Brian was around three years old, his dad decided to design and build a small train that he could actually ride in. Figure A shows a picture of Brian as a kid and his dad with that train. In true maker form, his dad located the materials to build it by scrounging around his workshop, visiting local junkyards, and perusing hardware stores.

If you look closely at the train's boiler, you may notice that it is a trash can turned on its side. The smokestack, drive cylinders, and cowcatcher structure were made from PVC pipe, and the cab and wheels were made of plywood that was scrounged from the workshop. A motorcycle battery powered a windshield wiper motor taken from an old Pontiac. The track was also made from readily available, inexpensive materials: The rails were made of PVC pipe and the cross ties were lengths of 2"×2" lumber. The only component Brian's dad purchased was the brass bell in front of the cab (though the bell's mounting bracket was certainly homemade). Through the years, Brian and his dad made many projects together. But the train was the first and the main one that instilled a lifelong desire in Brian to make and to share his creations.

When Brian's son, Evan, was around three years old, Brian wanted to make him something meaningful, like the train his dad had built for him. He wanted to pass down this passion for making to his own son, and his first thought was to make a train for Evan. He started drawing sketches with pencil and paper which quickly evolved into 3D models created with a CAD program. He began to acquire components and materials for the train and was almost ready to begin construction when suddenly everything was derailed (pun intended) by a "**maker moment**" – a sudden desire to experiment, to explore, and to play. What happened?

A few weeks earlier Brian had salvaged and brought home from work a roll of used conveyor belt or chain. He rediscovered it while putting components for the train under his workbench. When he saw the chain, he immediately thought about how it could be turned into a tank tread or caterpillar track. He set out a length of the

A   Brian and his father with the homemade train

B   Brian Bunnell and his son, Evan, in the Tot-Size Tank

chain on his workbench and contemplated how to support the ends of the chain while simultaneously driving and guiding it. Soon he was cobbling together a backbone frame with bearings and sprockets to engage the chain at both ends and built a track system he could drive using a hand drill. The train was forgotten as Brian began to imagine a kid-sized tank that Evan could ride in instead.

The tracked vehicle became known as the "Tot-Size Tank," as seen in Figure B. Based on the encouragement and recommendation of a friend and fellow maker, Tom Heck,

Brian submitted a write-up on the build of the Tot-Size Tank to *Make:* magazine. *Make:* graciously published the entire article on its website makezine.com and an abridged version of the article made it into the print edition, *Make: Volume 51*.

The issue of *Make:* magazine where the Tot-Size Tank appeared

INTRODUCING THE ADULT-SIZE TANK

Samer Najia has always been fascinated by tracked armored vehicles, especially those from World War II. As a boy he dreamed of creating an all-terrain tracked vehicle that could go everywhere. Samer grew up, went to engineering school, built and worked on several vehicles, and continued to dream about that tracked vehicle. He designed a variety of track systems and liberally modified designs of others, but nothing really gripped him.

Then one day, Samer happened across Brian's Tot-Size Tank online. He realized Brian had hit upon something he'd never thought of: using a food conveyor belt for the tracks!

One of Samer's go-karts looked like a candidate to become a tracked vehicle. He took a shot in the dark and asked Brian a question about the tracks in the comments section of his makezine.com article. Brian found the name of the manufacturer embossed on the links of the conveyor belt and responded to Samer, who bought some belts from the manufacturer and converted his go-kart first into a half-track and later into a tank big enough for him to ride around in himself (see Figure Ⓐ). The back-and-forth conversations between the two men led to a friendship, and Samer's

Ⓐ Samer Najia's Adult-Size Tank

Samer Najia on board an airplane he flies

own creation ended up on the *Make:* website, as well as on Hackaday (hackaday.io/project/20432-gokart-tank).

The global maker community is made up of thousands of people in all walks of life who love to share their ideas, show off their projects, and help each other out. Brian and Samer (that's us, your authors!) are continuing the partnership that began with tracked vehicles by collaborating on this book. The Tot-Sized Tank and the Adult-Size Tank will be used as examples throughout the rest of the book to illustrate mechanical engineering concepts and show how you can use those concepts in your own projects. While we don't expect you to build a tank, you're welcome to go back to those original articles and give it a try!

You can read the original articles here:
- makezine.com/projects/build-mini-tank-tread-electric-car/
- makezine.com/2017/04/05/gokart-tank/

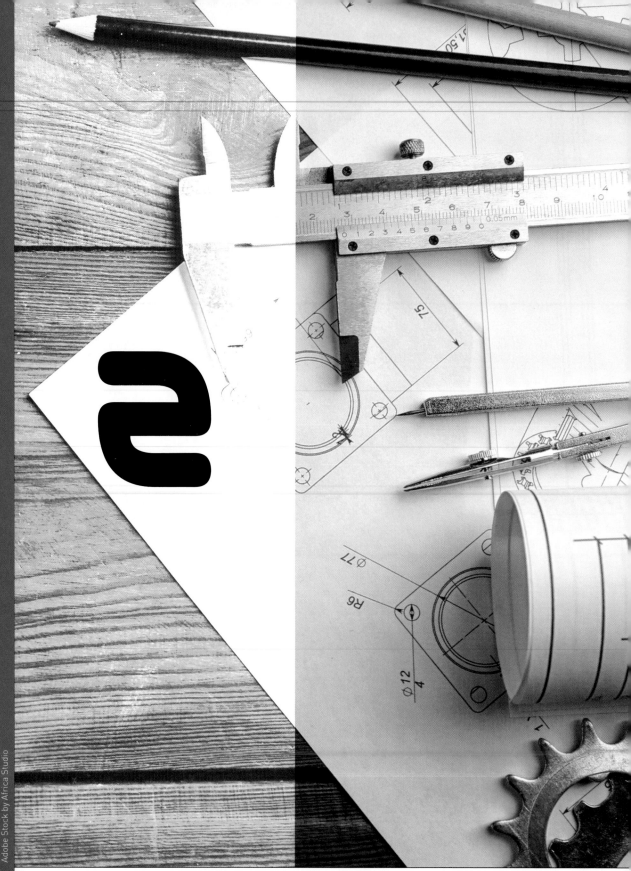

As you become a practiced maker, the act of creating things becomes more than just putting things together. Making becomes an expression, and a pursuit of your ideas, passions, and curiosities. We makers love to share our ideas, projects, and experiences with others. But before we can create and share our ideas, we must assimilate our thoughts, learn the skills and principles that are required, and work to make our ideas real.

In this chapter, we will walk you through a process that we have found helps us take our ideas from vague thoughts to fully functional realities. This process is not set in stone; it can and should be adapted to specific situations as needed. But it is a good path to follow when thinking through a project.

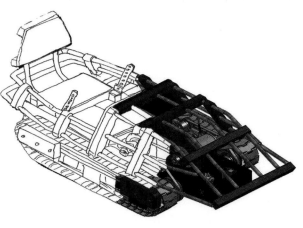

Tot-Size Tank - sketch to solid model

Project Realization Process

In outline form, here are the steps for fashioning your initial glimmer of an idea into a real-world mechanism:

- Define the project idea
- Define the specifications
- Conceptualize (sketch) the concept
- Perform the research
  - See what's already out there
  - Confer with others
  - Take a walkabout
- Fabricate the project idea
- Test and refine the project idea

Remember that these steps are not a hard and fast rules. The process can (and should) be quite fluid. Some of the steps can be in a different order, or the process might loop back onto previous steps as your project concept evolves. For example, you might do the research, at least partially, before sketching out your concept. And, it's not unusual for the concept to change dramatically after conferring with other people, requiring backtracking in the process. This is not a rigid set of rules; the creativity of making includes being creative with the process, too!

Define the Project Idea

What are you are setting out to make? Why do you want to make it? What is the purpose of making this thing? These are all good questions to ask at the beginning of a project. Answering these questions will help refine initially vague ideas about a particular project. In most cases, these questions will come quite naturally, though they are nonetheless important to ask and answer. At this point, you are starting to think about the initial scope of the project.

For mechanical engineers, this is a very important first step in any project. We ask general questions about what it is that we need to design. This thought process then naturally flows into *why* this design or project is needed. In the realm of formal mechanical engineering projects, there is always a specific, cost-justified reason to design and make something. This formal project may be a complete machine, a new feature for an existing machine or even something for a research and development effort. In a business setting, there will always be business or cost-related justification for a project that will, in some way, make money now or later for the business.

Interestingly, this is one area where formal Mechanical Engineering and mechanical engineering for makers can be quite different. As makers, most of the projects that we have made (and will make) are justified without any thought of financial gain or even financial justification. In many cases, the answer to the questions, "Why do you want to make it?" and "What is the purpose of making this thing?" may simply be, "I want to see if I can." That is a perfectly acceptable reason for a maker to embark on a particular project!

A few years ago, Brian had been given a 9,000-volt neon sign transformer (Figure Ⓐ). Wondering what he could make with it, he quickly honed in on the idea of trying to build a Jacob's ladder using the transformer as the power source.

There was no real reason to try this; Brian simply wanted to see if he could make it work. (He also thought that it would be pretty cool to display vast amounts of arcing electrical energy traversing up two bars, and jumping through thin air!) Well, it DID work, and Figure Ⓑ shows the resulting project. Figure Ⓒ is a long exposure of the Jacob's ladder in action. This project, done with no more

What is a Jacob's Ladder?

A Jacob's ladder is a type of spark generator. It is made up of two vertical rods that are slightly diverging from one another. In other words, the rods start out closer at the bottom then get further apart at the top. A large potential or voltage is passed through the air between the rods in the form of a spark or arc. This arc then travels up the rods until the gap is too large for the potential to maintain the arc, where the arc "breaks" or stops. The process then starts over with another arc formed at the bottom of the rods. As long as the transformer continues to supply potential, the arc will continue to form, rise up the rods and eventually stop, over and over again.

 The Jacob's Ladder old neon sign transformer

reason than to see if he could, ended up being a favorite at science displays and Maker Faires!

Define the Specifications

Once you understand the initial idea of the project that you wish to make, you will need to dive deeper into the specific aspects or specifications of the project. Think about what you want your project to do, and how you can make that happen.

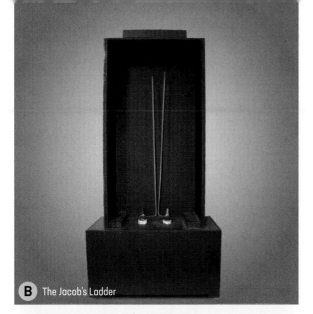

B The Jacob's Ladder

Take, for instance, the Tot-Size Tank project as introduced at the end of the previous chapter. Early on in the project, Brian had to nail down quite a few parameters or specifications that served to hone in on and guide the design and construction of the tank. Brian decided that the Tank must:

- Be the correct size for his son to ride and operate
- Be simple enough for a child to drive
- Have enough power to propel itself and his son
- Be safe for his son and bystanders
- Be better and faster than other power-riding toys (goal: 12mph)
- Be budget-friendly (mostly use stuff he already had)
- Be built with tools he already had
- Store easily
- "Grow" with his son
- Look cool!

The Adult-Size Tank, also introduced in the previous chapter, had to have its requirements defined, too. Samer decided that the Tank must:

- Operate at a specific speed (5mph top speed)
- Traverse certain sized obstacles (able to climb an obstacle 4" tall)
- Have differential steering by adjusting the power of each set of tracks (versus differential braking, where one side is stopped)
- Allow for the option of reversing each track separately
- Be light enough to be carried/maneuvered by one person

As you can see from these two examples, project specifications should include performance expectations in addition to functional characteristics (the machine should do X or Y). For example, specifying the anticipated performance characteristics will mean the difference between building a race car and a pedal car. Performance expectations can sometimes seem unattainable for a maker, but do not let that scare you off your project or limit your creativity. Engineers (and makers) can usually find clever ways around performance requirements to maximize strength, capacity, speed and other details with limited impacts on weight and size. With a little creativity in your design, you will find the balance that meets your project expectations.

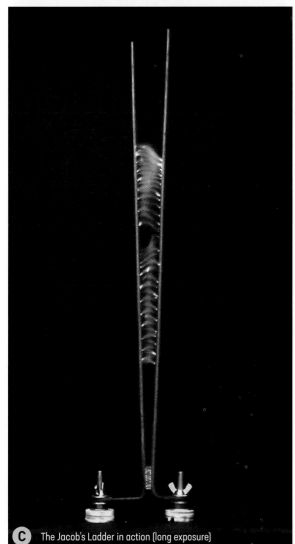

C The Jacob's Ladder in action (long exposure)

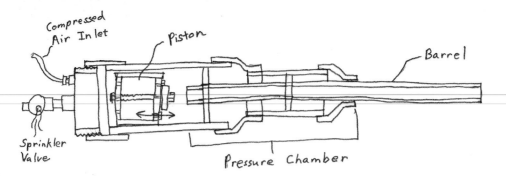

A Sketch of pneumatic (air) cannon concept

Pneumatic Air Cannon
(Cross-Section View)

Making a list of specifications early on in your project will help refine your ideas, hone in on your design, and maintain scope. Larger projects can assume a life of their own, growing in complexity well beyond initial intentions. Sometimes this can be a good thing, but most of the time, unchecked growth can lead to problems. The project can become too complicated, making it technically infeasible or too costly. Either situation can lead to an unsuccessful project: You might get fed up with it and abandon it, or it might simply get too expensive to continue.

Conceptualize (Sketch) the Concept

ROUGH SKETCHES

Before mechanical engineers ever start building anything, we spend a significant amount of time visualizing the project. One of the most effective ways to extract project ideas from the many scattered and jumbled thoughts in our brains is to simply draw it. Sketching forces our project-specific thoughts and ideas into a more cohesive and practical object, device or mechanism.

So, irrespective of what it is you are making, go get a piece of paper and draw it. Your drawings don't have to be on engineering graph paper or even a clean sheet of paper. Sometimes we find ourselves sketching on napkins, receipts, or backs of used envelopes – anything will work! Don't worry if you are not an accomplished artist when sketching your project ideas; some of us make drawings that no one else can understand. As long as you understand it, that's okay.

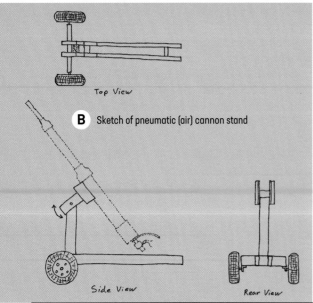

B Sketch of pneumatic (air) cannon stand

Top View

Side View

Rear View

Figure **A** is an example of a rough sketch of a pneumatic (air) powered cannon, also known as a potato gun. This sketch (originally drawn on scrap notebook paper) shows a side, cross-section view where you can see the planned inner workings of the cannon.

MULTIPLE ANGLES

Drawing the project concept from a number of angles (i.e., points of view) helps you visualize and further refine the idea. More often than not, you will need to look at it from one or more of the following views: top, side, back, front, and bottom. Drawing multiple views can help determine where all the bits and pieces go, and how they all fit together (or run into each other).

As an example, Figure **B** is a sketch of a rolling stand on which to mount the pneumatic cannon. It was helpful to show the stand using a top, side, and back view to fully illustrate what the stand might look like. Also, a ghosted, dotted-line image of the cannon was included in just the side view to show how the stand might possibly hold the cannon. From these sketches, a finished project was realized, and Figure **C** shows the actual cannon mounted on its stand.

It may be helpful to draw individual parts and indicate the direction and range of movement of the parts on your drawing. If a lever moves up and down, will it interfere with another structural part, or something else that moves? Will a motor or wheel scrape against anything when a force is applied in some direction? Will you be able to align the pieces accurately using your design? All these factors can alter a design, add weight, or present other problems that were not initially apparent, but may become apparent when you make a sketch.

SUBSYSTEMS

Once an overall project is broken down into its individual parts, imagine how they all fit together and how they come together. The process can determine where things like fasteners (screws, nuts, rivets, etc.) go, and might also dictate when and how you drill your holes, as well as whether you get a pre-made part, fabricate it, or 3D print it.

When we were designing and building our tanks, we had a pretty good idea of what the end result might look like, but we took the time to do deep dives into each sub system (steering, propulsion, wheels, and so on), flipping the design of each part over and over in our heads, on paper, and in CAD. Figure **D** illustrates an initial deep dive into the drive concept for the Tot-Size Tank. This concept was best illustrated using top and side views. (We will be covering specific details of how this drive system

C Completed pneumatic (air) cannon on its stand

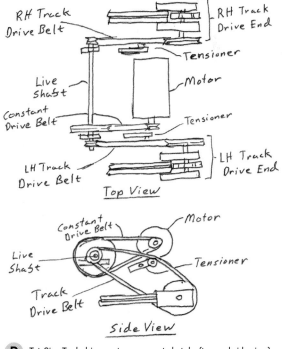

D Tot-Size Tank drive system concept sketchs (top and side view)

A   Tot-Size Tank drive system (top view)

B   Tot-Size Tank drive system (side view)

was designed and how it works in subsequent chapters.) Figures **A** and **B** are photos showing the actual Tot-Size Tank drive system, taken from the top and side, respectively. You can see how the sketches and photos look similar to each other.

SCALED DRAWINGS

Once your rough project sketch is complete, take the time to develop your sketch into an actual drawing, with dimensions and parts to scale and placed where expected in the final design. It is difficult to scale a design correctly in a sketch, so dimensioned drawings will help identify potential fit problems early on, as well as clarify the dimensions of the finished design.

DIMENSIONAL DRAWINGS

Finally, it may be beneficial to take the time to illustrate the design in three dimensions. Imagine it moving in 3D, and look it over again and again from all angles (top, side, back, front, and bottom). Look for potential areas of obstruction. Try to identify where to put primary fasteners (those that are critical to the assembly) and make sure they don't interfere with other parts. Imagine the moving parts, and make sure that these parts will be able to move freely. If applicable, picture the person driving or using the object, and see if there is enough room for them to move as needed.

Drawing each angle and individual part or system may seem daunting, or complicated, but really it isn't that hard. The work that you put into it will definitely pay dividends in the end. The more you decompose whatever you are building to its individual sections, and then visit and revisit the overall assembly, the more likely your end result will turn out and function as you intend.

Perform the Research
SEE WHAT'S ALREADY OUT THERE

Many makers start refining their project ideas by looking at what others have already done. Take advantage of Maker Faires, go to local science and/or artisan shows, and visit areas that have examples of the types of things that you are interested in making.

And, of course, do some online research. A lot of ideas have already been vetted by others, and it never hurts to see the approaches others have taken to solve a particular problem. Sometimes, just seeing how someone else solved an issue (even if isn't the exact issue you are interested in) can open your eyes to new possibilities for your own project concerns. Roaming through the maker-related sites — Makezine, YouMagine, Thingiverse, and Instructables, to name a few — can really get the problem-solving and creative juices flowing! If you don't mind revealing your own ideas to the world, use these same sites to post your musings and see what feedback you get. As you will recall, we (your authors) actually met this way!

CONFER WITH OTHERS

After sketching the project from a few angles, and working out some of the obvious issues that the sketching reveals, float it past a friend or two. Talking your project ideas through with others can help pull together the ideas into a more practical, cohesive concept.

Talking with others gives you different perspectives; your friend will see your project through their own lens, ground in their own unique experiences, interests, and education. Being open to their feedback (both positive and constructive) can provide you great insight for improvements and changes to your project.

Oddly enough, some of the best critiques actually come from kids. A child's insight can be unbiased, innocent, and honest. Children see aspects of project ideas that may not be obvious to adults, and their feedback can spark further development or simplification of the project. Children can be more creative than adults, because they are less encumbered by life's experiences. Try explaining your project details in words and concepts that a child will understand; this exercise will force you to simplify your own thought processes. Let's face it: the KISS (Keep It So Simple) philosophy is usually the best approach!

TAKE A WALKABOUT

As a project idea matures, you may reach a point where you need to look at, touch, and even play with materials or items that you might use to build the project. This is an opportune time to take a walk through your personal workshop, or a local hardware store, and see what materials are there and how they might be adapted for the project. We like to refer to this as "going on a walkabout."

During this process, look for structural materials (wood, metal, plastic, angle, sheet, and tubing), fasteners (screws, nuts, bolts, rivets, threaded rods), composites (fiberglass sheet and epoxy), and wheels and bearings (as you might find for doors, dollies and so on). Look at how long or how wide, how strong or how heavy these things are and decide if you can use them, with or without modification. Basically, look at these materials with your project specifications in mind, and determine if using them could save you some trouble, or add functionality you had not anticipated. Also, when going on your walkabouts, bring your sketches, sketch pad, and pencil. You may find yourself modifying those initial sketches when you see what is available. A walkabout may, and most likely will, spark new ideas that you will want to record.

Fabricate the Project Idea

An important and necessary part of the project realization process is to determine how you will make or fabricate the components and assemblies of the project. The fabrication process is critical to your design. You must take into account available materials, tools, skills, and time to build your project. Consider what machines and tools you have access to. If you determine that a component is too complicated for you to make with the tools that you have, consider purchasing it — or the specialty tool to make it (makers love new tools!).

We will be talking about fabricating through hands-on projects throughout this book. You will become familiar with many fabrication techniques using various tools and materials while completing these projects, so we won't get into detail here.

It's no good to design something cool, only to find that you can't really build it without more advanced tools or manufacturing processes. As perfected as an idea may seem, take the time to fabricate it in your mind. See yourself cutting the parts, drilling the holes, and tightening the nuts. Work through the assembly of every part, looking for potential problems and bottlenecks. More than likely, you will find improvements you can make or problems you did not originally foresee.

One way to refine a project design is by building a working miniature – this is called **scale modeling**. Scale modeling has obvious limitations, particularly where the finished project will be introduced to large forces that can't be simulated or the project materials are not available in small scale, but you can usually find suitable

Staying on Track: Rules of Thumb for Thinking Through and Designing Your Project

Remember, as you work your way through your design, try to do the following:

- Limit the number of parts as much as you can.

- Limit the number of moving parts as much as you can, since that reduces complexity.

- Try to use the same set of parts as often as you can in the project, since this simplifies fabrication. "Parts" might be individual parts, subsystems, or subcomponents.

- Try to use as many off-the-shelf parts as you can. Not too many of us have a full machine shop, access to heavy-duty tools like a lathe or table saw, or other implements. If you can use standardized parts and standard lengths/widths of those parts (that is, if you don't need to cut, or can design around standard sizes), do so.

- Look for alternatives and ideas done by others and do your research, especially for more complex projects. If you can, get your hands on any plans or blueprints you can and consider whether adopting (or adapting) someone else's design to your purposes may be a better alternative than a "clean-sheet" project.

- Makers should live by the adage, "Measure twice and cut once." Remember that it is harder to put parts together (add material) than it is to cut them apart (remove material). We have ruined countless parts because, in our excitement to get moving, we didn't measure correctly, or made some mis-calculation. While we can understand that some of that is unavoidable, it really should not be the norm.

substitutions. As an alternative, look into 3D printing your parts, either yourself or through a service. Granted, this might get expensive (a good reason to get your own 3D printer), but you might find that having model parts in hand can prove to be invaluable in making decisions for fabricating the final parts.

The scale model in Figure **A** uses a chassis that is identical to the Adult-Size Tank. Samer added a 3D printed body shell for looks. Figure **B** shows yet another 3D printed, scaled model of a concept for tank treads and drive wheels.

When it is not practical to build a fully operational scale model, consider making one that addresses a single aspect of your design. When Samer was designing the Adult-Size Tank, he built several smaller track systems to test out various approaches in miniature. Even though the end result looked little like the operating models, the working mechanisms could be fleshed out in miniature with minimal expense. This is an example where scale modeling via 3D printing proved very beneficial to the project design.

Sometimes, projects lend themselves to **prototyping**, or creating initial, full scale (actual size) trial versions to test or learn from in anticipation of the final creation. No doubt you have seen products come out with version numbers associated with them, and many times, Version 1.0 is a prototype. You can use this concept in your own projects. Perhaps your grand idea could use a little adaptation and modification as you go, so that by the time the project is complete, most if not all of your design problems have been addressed.

For example, while Samer was developing the Adult-Size Tank, he was unable to locate a suitable track to use. Everything he found in his research did not meet one criterion or another, so he decided to make the tracks himself. The specifications for the tracks indicated that they needed to be durable, flexible, and able to fit the tank drive wheels and sprockets.

His earliest design was composed fully of metal parts. After making one sample, it was determined that 100 links would be needed, but the time needed to make and assemble all 100 would be .prohibitive. Further, the overall weight of the tracks alone would easily overwhelm

A Early tank scale model (3D printed) for testing differential steering

B Tank treads and wheels (3D printed)

C Tot-Size Tank conveyor chain track material

D Machined wheel (drive sprocket)

the available power plant (a small gasoline motor). Going back to the drawing board, he then designed a hybrid metal/wood track, which turned out to be too heavy as well, followed by a resin-cast, and then an aluminum-cast version. (Also, the aluminum version would have required the consumption of large amounts of fizzy drinks to harvest the aluminum cans.) In the end, none of these solutions were feasible. Finally, all of the prototypes and models were scrapped and the tracks were simply purchased.

As "brilliant" and creative as the track link designs may have been, modeling and prototyping led to actually choosing against those options. Samer's final decision to purchase a pre-made product (the same conveyor chain used as tracks for the Tot-Size Tank, as shown in Figure **C**) proved much more economical and very effective. This is a good example of where prototyping and collaboration with other makers can be indispensable.

A final note on fabrication is to consider outsourcing. Your time is valuable and sometimes it is easier and more cost-effective to have a service do some basic fabrication for you. Samer, for example, found it cheaper and easier to have some of the wheels on the Adult-Size

Tank machined by a service that accepted a CAD drawing. Because of that service's machinery, they were able to deliver multiple wheels (18 in all) for a far lower cost and more quickly than hand-cutting each wheel. Figure **D** shows one of these machined Adult-Size Tank wheels. Additionally, if part fabrication will be complicated, or they need to be precise, it may be worthwhile to have a CNC machine make the cuts or drill the holes. As a bonus, if you ever need to regenerate the part, you can always use the original as a template to do it yourself or simply order a replacement.

Other makers are also great resources for outsourcing manufacturing. For example, if another maker has a 3D printer, and you have the ability to weld metal, you can trade your tools and skills to help each other with your projects, or collaborate on one together. With the recent proliferation of CNC routers, desktop lasers, vinyl cutters, 3D printers, and other tools, it makes sense to leverage the tools and skills other makers have when you don't have the time, space, or funds to obtain them yourself. You could expand on this idea to include local maker guilds, woodworking clubs, etc. where you can meet other makers. As a bonus, many of these organizations have materials available and tools for rent.

Tracking Further: Fabricating and Prototyping Track Links for the Adult-Size Tank Version 2.0

Even though both the Tot-Size Tank and the Adult-Size Tank used purchased tracks, the limitations of these tracks became evident when trying to scale the design and add functionality. The tracks we purchased were molded with a small arrow on the inside surface. The arrow (or conveyor direction) wasn't an issue until Samer added the ability to reverse the direction of the tracks on one side of the Adult-Size Tank. When the track ran in reverse, it promptly got "thrown" off the sprockets, immobilizing the tank.

Since Samer had added "reverse" as a new requirement for the Adult-Size Tank (version 2.0), this issue prompted him to conceptualize, design and prototype new tracks.

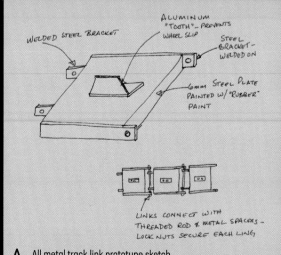

WELDED STEEL BRACKET

ALUMINUM "TOOTH" - PREVENTS WHEEL SLIP

STEEL BRACKET - WELDED ON

6mm STEEL PLATE PAINTED W/ "RUBBER" PAINT

LINKS CONNECT WITH THREADED ROD & METAL SPACERS - LOCK NUTS SECURE EACH LINK

A   All metal track link prototype sketch

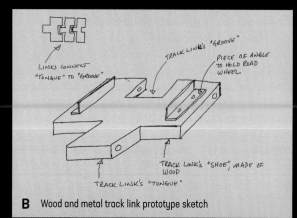

LINKS CONNECT "TONGUE" TO "GROOVE"

TRACK LINK'S "GROOVE"

PIECE OF ANGLE TO HOLD ROAD WHEEL

TRACK LINK'S "SHOE", MADE OF WOOD

TRACK LINK'S "TONGUE"

B   Wood and metal track link prototype sketch

The all-metal track and the hybrid metal-and-wood track (Figures **A** and **B**) were both scaled-down prototypes. Even though these track prototypes "worked," they weren't feasible designs for a number of reasons.

First, they were just too heavy (remember, one specification for the original Adult-Size Tank was that it be light enough that Samer could carry/transport it himself). Also, this tank was designed to operate with a motor, having a fixed maximum power output, that Samer already had available. That motor dictated a lot of other criteria and requirements such as weight. Let's face it, tanks are heavy already, and there is no escaping that the tracks are going to add significant weight. So, the design would need to take this into account, so that the motor and tracks could work together to meet the overall requirements. (This is an example of one or more subsystems' specifications impacting another system or subsystem.)

Second, the fabrication process for the prototypes brought up a new requirement (i.e. a specification): The tank should be completed within a reasonable timeframe. The time involved in making and assembling the all-metal and hybrid metal-and-wood tracks was just too much.

These discoveries via our prototypes caused Samer to revisit (and update) the project criteria specifically regarding the tracks for this Adult-Size Tank version 2.0:

- Reasonably light track links (of course, "reasonable" leaves some ambiguity, so let's say that the entire track must not weigh more than 30lbs per side, This factors in the weight of the rider, the vehicle, and the power from the motor/engine available
- They must be easily replaced
- They must be able to run on smooth surfaces as well as rough ones (this just means they have to either be rubber coated or have a rubber "shoe")
- They must be able to run in reverse
- Each link must be composed of 5 parts or fewer (100 track links means 500 parts)
- They must be resistant to water/corrosion and wear (which in this case, means adding the rubber "shoe" we have already mentioned)
- They must support a road wheel that is 50mm wide, and have a method to prevent a track from sliding off the road wheels

Additionally, because the length of each link dictates the overall length of the track, which then dictates (to some degree) the final length of the vehicle, we need "x" track links to support "y" weight and carry a person at most "z" long, which translates to a track of between "A" and "B", which then translates to "W" track links. See the dependencies? And none of these were listed in the original requirements for the track! So that's a new, unforeseen requirement that had to be added.

An initial prototype track link is depicted in Figure C. Keep in mind that fabricating in miniature can be vastly different from fabricating at full scale. The materials and ability to machine them (including the tools) change with size. These links could have been cut by hand, but since he had a 3D printer available, Samer printed the 5 or 10 links needed for testing. He printed them at half and then full scale to see the effects.

As you can see in the 3D-printing software image of the track link (Figure C), the track design was modified from the original prototyping sketches. The newer design provided a "dam" to prevent the track link from sliding off the road wheels, with a hole in the middle of the link to support a sprocket tooth for the tank's drive sprocket.

Figure D shows the track printed as a half-scale model to test its functionality prior to printing at full size for more extensive testing. Typically, a scale model will behave similarly to a scaled-up version. So, according to this scale model "test," the links seemed to articulate (i.e., move) well, mesh well, and are very solid.

This design ultimately did not work, because the axle on which the wheel was mounted interfered with the track's outer ridges. But prototyping helped refine the design early enough in the process so that Samer did not have to go long down a path that would have required a complete revisit later.

Figure E shows a final design for the track with a road wheel sample in place. While prototyping proved invaluable in the process of identifying, validating, and testing a design, Samer settled on two rows of the original track with a piece of material in between cut to the right length to create a 'shoe'. If the link is printed, that material will simply be printed as part of the link. If Samer uses two of the original track links, he could

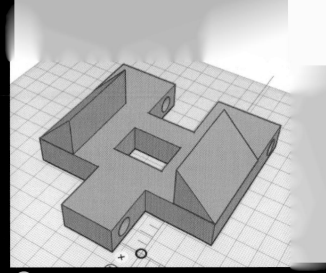

C Adult-Size Tank track link prototype (3D view)

D Adult-Size Tank Track link samples at 50% scale

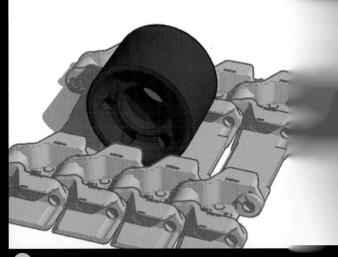

E Final tank tread design for Adult-Size Tank V2

bolt on a piece of aluminum plate. The possibilities are endless. Also, this design helped Samer determine the overall dimensions of the track, as well as the number of links that were required.

Test and Refine the Project Idea

As you already know, engineering sometimes involves compromises, and these compromises may bubble to the top when you conduct your tests. Perhaps you have made some assumptions or set some specifications that are not really compatible with the design. When you start testing, your invention may break when you least expect it, or not function as expected. There is nothing wrong with this. It's part of the process. Do not be afraid to fail. A maxim all makers should become familiar with is, "failure breeds success," meaning you cannot expect to succeed the first time every time. As one of our instructors used to say, "If you have not fallen, then you have not learned anything."

When you test your design, try to test individual assemblies before you test the whole system. When Brian was working on the Tot-Size Tank, he separated the major components into discrete systems and was able to test each individually. For example, he connected a hand drill to the tank's drive wheel and test and refine the track system (Figure **A**).

If you wait until the entire project is finished before testing, it may be too late to undo or redo something that isn't quite right. If you think of your creation as a system of subsystems, you will be able to be more flexible and find early indications of problems you might not otherwise encounter when the overall project is complete, at which time it may be too late.

A Using a hand drill to drive a tank track

Now, we will take the project realization process that we just discussed and put it into practice by making our very own, obnoxious-to-the-neighbors air horn! Each process step shown throughout the project has been "filled in" as an example for you to follow, but feel free to practice the process by filling in your own answers.

DEFINE THE PROJECT IDEA

What do you want to make, and why do you want to make it?

We want to make an Air Horn to illustrate the teachings of this chapter. We also want this project to be fun for you, our readers, to build. As an added bonus, we want to see how loud this thing can be!

DEFINE THE SPECIFICATIONS

What specific things are important about this project?

The Air Horn must be:
- Be inexpensive to build
- Be built from readily available materials
- Be simple to build using common tools
- Be "powered" by a common air mattress pump
- Not take more than a few hours to make
- BE LOUD!

CONCEPTUALIZE (SKETCH) THE CONCEPT

Figure **B** shows an initial sketch of a single air horn.

PERFORM THE RESEARCH

See What's Already Out There

Here are a few other air horns that we discovered on the internet:
- makezine.com/2011/07/01/how-to-hand-pumped-pvc-foghorn/
- instructables.com/id/Easy-PVC-Fog-Horn/
- brilliantdiy.com/video-diy-your-own-120db-worlds-loudest-homemade-air-horn-requiring-only-some-basic-materials/

Confer With Others

We reached out to friends and family to get their input on this specific project idea:
- My son said, "Why is there only one horn? Wouldn't two be louder?"
- My wife said, "Nice. Can I use this to wake the kids up in the morning? (Laugh) Also, please make sure that it's easy to store. We have too many contraptions out already."
- My co-workers said, "Cool, Man! How loud will that thing be?!"

• • • Take a Walkabout • • •

As you can see from the sketch in Figure B, the air horn is conceptually simple. But, we need to determine the items that can actually be used to make it. As mentioned before,

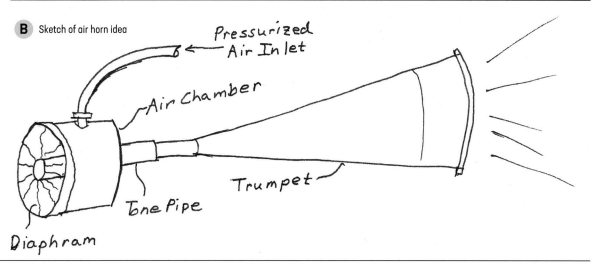

B Sketch of air horn idea

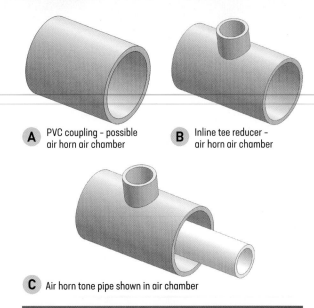

A PVC coupling - possible air horn air chamber

B Inline tee reducer - air horn air chamber

C Air horn tone pipe shown in air chamber

How an air horn works

An air horn is a simple pneumatic (air) machine. The image below is a section-view sketch showing the internal structure of a simple air horn. Pressurized air is introduced through the air inlet to the air chamber, causing the pressure in the air chamber to rise. The rear of the air chamber has a thin, flexible, plastic diaphragm that seals the end of it. Concentric to the air chamber, going through the middle of it, is the tone pipe. The end of the tone pipe, protruding through the air chamber to the left, is also initially sealed by the diaphragm. However, as the air pressure in the air chamber builds, the diaphragm stretches out slightly, breaking the seal with the tone pipe, causing a very small amount of pressurized air to escape into the tone pipe. Then the elasticity (i.e., stretchiness) of the diaphragm causes it to move back against the tone pipe, thereby sealing it back off. This in-and-out movement of the diaphragm happens very rapidly, vibrating the air in the tone pipe, and the vibration of the air is what produces the sound of the horn. The tone pipe geometry directly influences the sound that the air horn will produce. As you make the tone pipe longer, the sound will be a lower pitch. This is due to the increase in the volume of air being vibrated within the tone pipe. The trumpet serves to amplify the sound emitted from the horn.

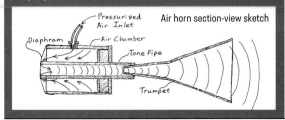

Air horn section-view sketch

Pressurized Air Inlet
Diaphram
Air Chamber
Tone Pipe
Trumpet

a good initial step is to visit a local home improvement or hardware store.

Below we imagine Brian's thought process as he goes on a walkabout for this air horn project and gravitates to the plumbing area, looking at the various PVC pipes and fittings. If you do not know where to start when contemplating components on a walkabout, start with the largest, most central, base component of your project concept.

Where shall I start? I'll start with the air chamber – after all, it is the central component of the air horn. It requires a primary, large inlet and outlet, and it also needs a smaller, secondary inlet for the pressurized air. With these attributes in mind, let's start looking at the fittings that will achieve most or all of these requirements. If most, but not all, of the requirements are achieved by a specific fitting, maybe we can add the missing element by simple modification.

First, I'll grab a 1½" PVC pipe coupling (Figure **A** *). It looks about the right size, and it has a nice opening on each end to receive other fittings. The sockets can serve as two out of three openings required for this part of the air horn: the main large inlet and main large outlet. However, there is no provision for the third requirement: the small inlet for the pressurized air. So, maybe I could modify the coupler fitting to accommodate the pressurized air inlet. I could drill a hole in the side of the coupling and tap the hole (cut screw-threads) with a pipe tap tool to receive a threaded airline fitting (See Tracking Further: Using Ingenuity – Self-Tapping PVC With a Metal Pipe Fitting on page 40).*

However, I'd like to find a fitting that geometrically achieves all three requirements. I'll go a bit further down the plumbing aisle, looking for something that may be able to achieve this goal. I'll grab another type of PVC pipe coupler called an inline tee reducer. It is like the coupler fitting, but also has a smaller, side-entry feature designed to accept ½" pipe. Eureka! This 1½" inline tee reducer (½") fitting, as shown in Figure **B** *, requires no modification. It can be used as is for the air chamber.*

Now, I need to find something that can be used as the tone pipe within this inline tee reducer fitting (i.e., air horn air chamber). As seen in the sketch, the tone pipe must be smaller than the large inlet/outlet of the air chamber, and longer than the length of the air chamber. Also, it must be able to be mechanically secured within the air chamber. So, what about a short length of PVC pipe? I'll go over to the 2-foot length PVC pipe display where there are several diameters to choose from. Since the air chamber is 1½" diameter, I'll start with a ½" pipe. Hmm. When I compare it to the air chamber fitting, it looks too small, so I'll go up to a ¾" diameter pipe. This pipe diameter looks just right when placed within the air chamber fitting, as shown in Figure **C**. We are making progress!

Now, how do I mechanically secure the tone pipe concentrically within the air chamber, without touching the sides? There is a 1½" reducer bushing near where I found the air chamber fitting, and it fits the air chamber fitting perfectly (Figure **D**). Also, it has an interior hole to accept the ¾" PVC tone pipe. But wait: There is a ridge inside the reducer that is intended to prevent pipes from passing all the way through. That's a problem, since the design of the air horn requires that the ¾" tone pipe passes completely through the reducer as shown in Figure **E**. I guess a little modification will be required once I get it home. (More on this later.)

I'm pretty sure that I have something to use as a diaphragm (simply thin plastic film, like that used to make sandwich bags) at home already, but I need to find something to secure the diaphragm to the air chamber. In the plumbing section of the store, there are worm-screw, bands-style hose clamps that look perfect for this application (Figure **F**).

The largest hose-clamp that this particular store has only goes up to 1¾" diameter, and I know that this is not large enough. (Side note... On these walkabouts, it is important to carry a small tape measure with you. It comes in handy in situations like this, where you need to check the outside diameter of the air chamber fitting. If you forget to bring one with you just borrow one from elsewhere in the store. Pretty much any store where a maker would go on a walkabout will have tape measures.)

The outside diameter of the air chamber fitting is around 2³⁄₁₆", so the largest hose clamp at this store will not work. I guess I get to take my walkabout to another store for this particular part. Oh well, that's not uncommon. It usually takes visits to more than one store to take a project from concept to something actually makeable.

But, where do I go next? Where can I find a hose-clamp large

Staying on Track: Following Your Intuition

You will, in time, come to rely considerably on the intuition that is built on your experience and that of others. In the case of the tone pipe, the initial size could have worked per the design, but it just looked too small. Intuition suggested going up a size, which proved the right decision. Not only did it look better, it turned out that there was a reducer fitting available that allowed for this tone pipe size to be secured within the fitting chosen for the air chamber.

Even in formal mechanical engineering design work, mechanical engineers rely heavily on experience and intuition. It is simply not practical to analyze every aspect of a design with calculations. We find that the only time, as mechanical engineers, that we use actual theory/math is when some aspect of the design is critical or our intuition tells us that there could be a specific problem with an aspect of a design. So, trust and run with what your intuition tells you! It will only get better with time and practice as a maker!

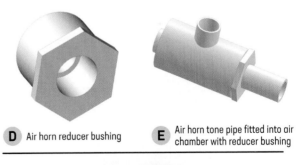

D Air horn reducer bushing

E Air horn tone pipe fitted into air chamber with reducer bushing

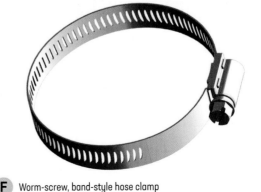

F Worm-screw, band-style hose clamp

(A) Air horn diaphragm shown fastened to assembly with pipe clamp

(B) Funnel used for air horn "trumpet"

(C) 3D design for the dual air horn project

(D) Parts required for dual air horn project

(E) Dual Air Horn Materials

» **PVC Pipe – ½" × 5' – standard schedule 40** (see cut list below), **1 length** (MMC* part # 48925K91)
  • Cut to 6" length, 2 pcs
  • Cut to 1½" length, 4 pcs
» **PVC Pipe – ¾" × 5" – standard schedule 40** (see cut list below), **1 length** (MMC part # 48925K92)
  • Cut to 6" length, 1 pc
  • Cut to 5" length, 1 pc
» **PVC Fittings:**
  • **Inline Tee Reducer – 1½" to ½", 2 pcs** (MMC part # 4880K506)
  • **Reducer Bushing – 11/2" to 3/4", 2 pcs** (MMC part # 4880K333)
  • **Side Outlet Elbow – ½", 2 pcs** (MMC part # 4880K631)
  • **Tee with ½" NPT Threaded Side Outlet – ½", 1 pc** (MMC part # 4880K41)
  • **Pipe Cap – ½", 2 pcs** (MMC part # 4880K510)
» **⅜" Diameter Tube Push Fit – ½" NPT Thread, 1pc** (MMC part # 9807K360)
» **Worm-Screw: Band-Style Hose (Pipe) Clamp – 2½", 2 pcs** (MMC part # 5415K190)
» **Thin Diaphragm Material** (we used plastic sandwich bags), **2 pcs**
» **Funnel** (we used different sized automotive oil-change funnels), **2 pcs**

* MMC = McMaster-Carr www.mcmaster.com

enough to go around the air chamber? Who would typically use large hose-clamps and why? Maybe people working on cars. There are certainly a plethora of hoses in a car, from fuel lines to brake lines to radiator hoses. So, it looks like the next stop is to the auto parts store.

The auto parts store has a rather large display of hose-clamps, of seemingly all sizes, and there, right on the peg in front of me, is a hose-clamp that goes up to 2½" in diameter. This will easily accommodate the diaphragm material thickness (very thin) and the 2³⁄₁₆" air chamber outer diameter. Figure (A) shows a 3D CAD model of the diaphragm clamped to the air chamber using a hose clamp.

So, what parts am I still missing? I need to find something to use as the trumpet or amplification cone, Figure (B). Conveniently, I'm still at the auto parts store, and the oil change supply area has several different funnels. Perfect! I'll need to run back out to the car to get the ¾" PVC pipe with which I will be making the tone pipe because the trumpet (or funnel) must fit well or be able to be modified to fit well in it. Now, I need to try the funnels to find one that fits best in the tone pipe. Even better, let's get a couple of different funnels and maybe make this a two-tone, dual air horn!

Wow, this has been a successful walkabout! Now, I'm ready to go home and make this thing!

The walkabout for the air horn project was successful! It resulted in gathering most of the supplies, but most importantly, it provided an avenue for the project to be thoroughly thought through. Now, it is ready to go from concept to reality.

FABRICATE THE PROJECT IDEA

After the walkabout (and being hounded by his son to add an extra horn), Brian modified the original design of a single air horn to a dual air horn. Figure (C) shows the updated project idea, modeled as a 3D figure. Figure (D) shows all of the parts that make up the dual air horn, which required another trip to the hardware store for the

F Exploded View of Dual Air Horn

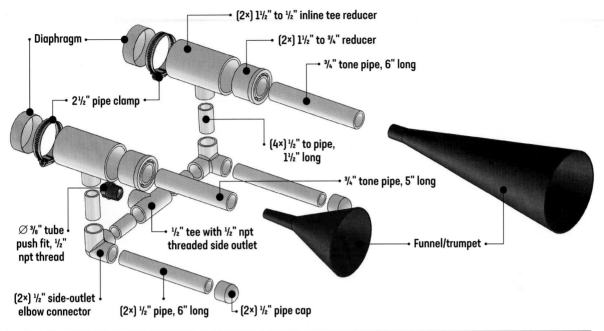

- Diaphragm
- (2×) 1½" to ½" inline tee reducer
- (2×) 1½" to ¾" reducer
- ¾" tone pipe, 6" long
- 2½" pipe clamp
- (4×) ½" to pipe, 1½" long
- ¾" tone pipe, 5" long
- Ø ⅜" tube push fit, ½" npt thread
- ½" tee with ½" npt threaded side outlet
- Funnel/trumpet
- (2×) ½" side-outlet elbow connector
- (2×) ½" pipe, 6" long
- (2×) ½" pipe cap

---

additional parts. These materials are referenced in the materials/cut list (Figure **E**), and each step in making, cutting and assembling these parts will be detailed in the fabrication steps to follow. Figure **F** is an exploded view of the dual air horn assembly detailing all of the individual parts. Where applicable, the cut lengths of PVC pipe used in the project is also listed in this figure.

Begin the fabrication process by cutting the PVC pipe lengths per the details in the materials/cut list. (If you need details on the best methods of cutting PVC, refer to the *Tracking Further: PVC Pipe Cutting Techniques* on page 34). Once you have the pieces cut, lay them out carefully and begin the assembly process, starting with the tone pipes and reducers.

The tone pipes for each side of the horn (made from ¾" PVC pipe) must be able to fully pass through the 1½" to ¾" reducer bushing fitting (also called a reducer). However, reducers are intentionally designed to prevent the inner pipe from passing completely through them. There is a small ridge inside the reducer that limits the smaller pipe to only be able to be pressed in a certain depth. PVC fittings are also slightly tapered in order to wedge together with a pipe or another fitting; this helps create

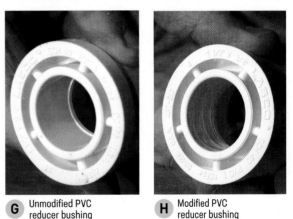

G Unmodified PVC reducer bushing

H Modified PVC reducer bushing

a seal between the mating surfaces. So, to allow the ¾" pipe to pass through, we will have to make some minor modifications to the reducer.

Figure **G** shows an unmodified reducer as it looks when purchased. If you look at the middle of the inside of the reducer, you can see the lip that needs to be removed. Figure **H** shows the same fitting after modification. A rotary tool (such as a Dremel) with a small sanding

Tracking Further: PVC Pipe Cutting Techniques

A   Ratchet pipe cutter & rotating pipe cutter

B   Tube cutter

C   Chop saw

D   Hand saw

E   Band saw

Adobe Stock - Владимир Солдатов

In making, there is almost always more than one way to do something. Take for instance the simple task of cutting a piece of PVC pipe to length. We can think of five different ways to complete this one simple task, all of which we have utilized at one time or another.

The best method for cutting PVC pipe is to use a plastic pipe cutter that works like a pair of scissors with a substantial mechanical advantage. There are two variants of the PVC pipe cutter (Figure Ⓐ). The ratchet pipe cutter is designed to cut pipe up to 1½" in diameter. The other type is a rotating pipe cutter, which is designed to cut smaller diameter pipe such as ½" and ¾". The downside to using these is that they can become quite tiring to the hand if you have more than just a few cuts to do or if you are trying to cut larger -diameter pipe.

Another way that we have cut PVC pipe is with a tube cutter (Figure Ⓑ). Although this tool is designed to cut metal tubing, such as copper pipe or metal electrical conduit, it works quite well with PVC pipe. Like the plastic pipe cutter, it is designed for cutting relatively small diameters up to about 1½". This cutter employs a small, sharpened wheel on one side and two flat rollers attached to a linear moving carriage on the other. The sharpened cutting wheel and the flat rollers trap the pipe or tubing to be cut. By rotating a screw threaded through the main body of the cutter and attached to the carriage, the rollers supported by the carriage exert a force on the pipe directly in line with the cutting wheel. As you rotate the pipe by hand, the cutting wheel cuts a V-shaped groove in the pipe. Once the sharpened wheel cuts to a given depth around the circumference of the pipe, you adjust the screw further in to apply force again to make the cutting wheel cut an even deeper grove. The pipe is cut completely through once the depth of the groove exceeds the wall thickness of the pipe. This tool works well for cutting PVC pipe; however, it is not the fastest method, and it does not give you a flat cut at the end of the pipe. The cut will have a tapered shape due to the V-shaped cutting wheel.

If we want to make a quick cut that results in a nice, straight, perpendicular end, then we typically use a miter or chop saw (Figure Ⓒ). A chop saw, will make a mess of fine PVC shavings. But it can be used to quickly make lots of cuts if you're working on a project that needs lots of pieces. Having said all of this, it is certainly the most dangerous approach to cutting PVC. We have found out the hard way that you should NEVER try to cut off a very short piece of pipe using a miter saw. That little piece of pipe can become a rather high-velocity missile when the blade catches it and flings it violently across the workshop! What constitutes a "short piece"? Good question! The pipe must have enough length on both sides of the blade so that the pipe is well supported by the rear support or fence. If this is not the case, then do not use a chop saw to cut the pipe.

An ordinary hand saw can also be used to cut PVC pipe (Figure Ⓓ). This does not make a nice, straight, perpendicular cut. Also, like the chop saw, it produces a lot of mess with fine, snow-like shavings. The advantage of using a hand saw for cutting PVC pipe is that it is very portable, inexpensive, and safe, compared to using stationary power tools like a chop saw.

And finally, we'll mention that it's possible to use a band saw to cut PVC pipe (Figure Ⓔ). We find that this works well for quick cuts if you are not too concerned about a nice, straight, perpendicular end. Because band saw blades are flexible, it can be somewhat difficult to get a nice square cut. It is also quite messy, in that the cut produces a lot of fine PVC shavings like the chop saw. Also, a band saw is a much more expensive and dangerous machine that not every maker will have or have access to. In other words, there are other ways to cut PVC pipe that are better, safer, and require less expensive tools, like the ones previously mentioned. However, if you happen to have a band saw, you can use it to quickly cut PVC pipe. In fact, if you have a project that requires PVC pipe to be cut in half along its longitudinal axis, this is really the best tool for the job.

The point to this discussion is not only to tell you how to cut PVC pipe, but also to drive home the concept that there are often multiple methods or tools that can accomplish a given task. Also, we want to illustrate the fact that you do not necessarily need to go out and buy an expensive tool or machine to accomplish a specific task. Use ingenuity. Be clever with what you have on hand to do the job.

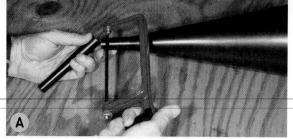

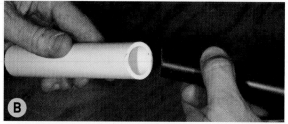

A

Cutting funnel spout to fit into tone pipe

B

Modified funnel ready to fit into tone pipe

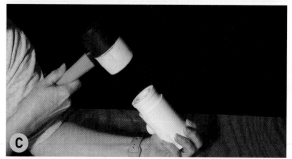

C

Using rubber mallet to 'seat' modified reducer in air chamber

D

Fitting tone pipe in air chamber assembly

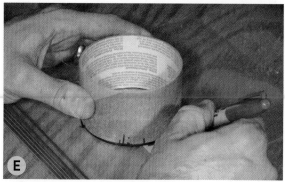

E

Using tape core as template to draw diaphragm circle

drum was used to carefully sand down the lip. The goal here is to remove the lip, allowing the tone pipe to completely pass through the reducer, while also achieving a reasonably airtight fit between the tone pipe and the reducer. You could also use sandpaper or a rounded file to do the same thing. As you are filing the lip down, be sure to test fit the ¾" pipe in the fitting. You want the pipe to slide completely through the fitting, but it should not be too loose. Again, you want a reasonably airtight fit. Repeat this process for both reducers

The trumpet for each horn needs to slip into the tone pipes securely. But neither should protrude too far into the pipes. The smaller, red funnel shown in Figure D on page 32 happened to fit perfectly in the tone pipe. However, the larger, black funnel went nearly through the entire length of the tone pipe. As you can see in Figure A, a fair amount of the funnel's spout end had to be cut off to make it fit.

To know where to cut on the spout, push the uncut funnel into the tone pipe until it wedges. Then mark the funnel with a permanent marker at the end of the pipe. Remove the funnel and measure down the spout 1" from the mark, and make a new mark – this is where you will want to cut. Now simply cut the funnel off at this position (Figure B). When inserted back into the tone pipe, it should go about 1" into the pipe.

Next, take one of the modified reducers and insert it into a 1½" to ½" inline tee reducer (i.e., air chamber). You may need to use a rubber mallet to tap the reducer into the tee to fully seat it (Figure C). After it is seated, slide one of the tone pipes through the reducer, but not all of the way through the tee (Figure D). Repeat this step for the other side of the air horn.

Now, make both diaphragms at once out of a single plastic sandwich bag. Draw a circle roughly 3¼" in diameter (a duct tape or masking tape core works well to trace around and get the appropriate diameter, as shown in Figure E). Then, cut through both layers of the bag, creating two thin plastic disks. Voila! Your diaphragms are complete!

Test-fit each diaphragm over the end of an inline tee that doesn't contain the tone pipe and reducer. Try to get it as concentric (evenly symmetrical and centered) as possible (Figure **F**).

Once in place, slip the hose clamp over the end of the tee, trapping the diaphragm between the clamp and the outside of the tee. It is very helpful to pre-adjust the clamp, without the diaphragm in place, such that it slips over the end of the tee with just a little bit of wiggle room. Make sure that the diaphragm is smooth and fairly tight across the opening of the tee (Figure **G**), and then tighten the clamp (Figure **H**). Repeat this step for the other air chamber assembly.

After the diaphragm is secure, carefully slide the tone pipe further into the air chamber until it touches the diaphragm. Then, continue to push it into the chamber, so that it slides roughly ¼" past the end of the tee. This will result in the slight stretching of the diaphragm, and is intended to create an airtight seal between the tone pipe and the diaphragm without ripping the diaphragm (Figure **I**).

At this point, both horn base assemblies should be complete, less the trumpets (Figure **J**). You will want to test each horn base assembly before moving on. First, clean off the outside of the ½" side outlet of the tee. Then, put your mouth on the side outlet and blow. You can generate enough pressure with your breath to make the horn sound. (Ours sounded like a sick duck without the trumpet, so if you want it to sound more like it will when complete, you can temporarily insert the trumpet into the tone pipe for testing.) If the horn does not generate a sound when you blow into it, blow harder. If there is still no sound, try carefully moving the tone pipe to stretch the diaphragm a bit more. If there is still no sound, check that the diaphragm is intact. Any small holes or rips will prevent the horn from working properly by allowing air to leak through the diaphragm. We found that these troubleshooting tips typically resolved any issues with the horn not creating a sound.

Next, you will need to make the base frame that also serves as an **air manifold** to supply both horns with pressurized air. A manifold is simply a chamber that allows for a single pressurized air inlet with multiple outlets. There is a single air inlet in the middle of the

Fitting diaphragm to air chamber assembly

Aligning diaphragm and hose clamp on air chamber assembly

Securing diaphragm to air chamber assembly with hose clamp

Adjusting tone pipe against diaphragm

Completed single air horn without trumpet

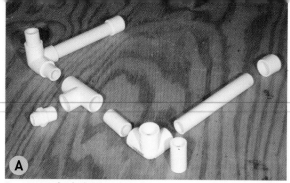

Components for dual air horn base frame

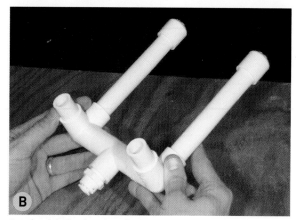

B

Completed dual air horn base frame

frame that will be used to supply the dual air horn with pressurized air. Figure **A** shows all of the components that make up the frame, in place and ready to be assembled.

Use Figure A and Figure F on pages 32 and 33 to gather the parts and assemble the frame, making sure that both "legs" (the two long, 6" lengths of ½" pipe) are parallel to each other, and to the tee outlet axis, prior to seating the pipes and fittings (Figure **B**). Again, use a rubber mallet to fully seat the pipes and fittings, making an airtight assembly (see *Staying on Track: PVC Pipe Friction Fitting* below).

You are almost there! Take each horn assembly and lightly slide it onto the base frame (Figure **C**). Before fully seating them to the frame, make sure that they are aligned parallel to each other and to the long legs of the base frame. Once in place, gently tap them with the rubber mallet to secure them to the frame (Figure **D**).

Lastly, slide the smaller funnel into the shorter of the

Staying on Track: PVC Pipe Friction Fitting

PVC pipe and its associated fittings are designed to fit together so that when they are cemented (actually solvent-welded) together, the joint becomes watertight. But cementing pipe and fittings together create a permanent bond, and that is not always necessary or preferred, especially for most of the projects in this book.

The outside surface of a PVC pipe's geometry/size is very tightly controlled and is consistent over the entire length of pipe. PVC pipe fittings have a very slight taper ending in a small ridge within the socket in which the pipe is inserted. This provides a relatively tight seal, even without cement. Connecting pipe and fittings together without cement is known as dry or friction fitting.

You will find that when you start to insert a pipe into a fitting it slides in fairly easily for about ¼"-⅜". As you continue to push the pipe into the fitting, you will find that it gets progressively harder until you reach a point where you cannot push any further (approximately ¾" deep). To fully seat a pipe within its fitting (about ¾" deep), it must be inserted to the ridge in the fittings socket.

When dry or friction fitting, you will most likely not be able to apply enough force by hand to get the pipe fully seated. This is where a rubber mallet comes in handy. With the pipe or fitting sitting securely on a hard surface, just give the other end a sharp whack with the mallet to seat the joint. However, if you need to align a fitting by rotating it, be sure to do this BEFORE you fully seat it. Once seated, there is a very large amount of friction between the pipe and fitting, making it quite hard to rotate the pieces relative to each other.

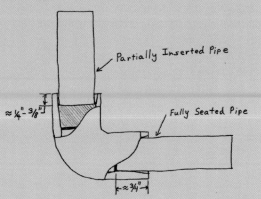

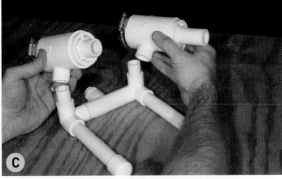

C Air chamber assemblies ready to be installed on base frame

D Completed dual air horn assembly without trumpets

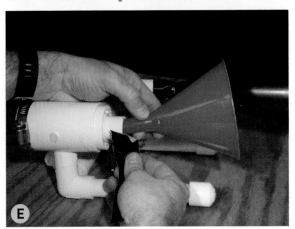

E Securing funnel 'trumpet' to tone pipe with tape

F Completed dual air horn assembly

two-tone pipes and the larger funnel into the longest tone pipe. Even if the funnels have a reasonably good friction fit with the tone pipes, you may still want to secure them with tape. A reasonably strong-yet-flexible tape works best, such as duct tape or electrical tape (Figure **E**).

Figure **F** shows the completed dual air horn.

TEST AND REFINE THE PROJECT IDEA
Now, it is time to test the entire system. A hand-operated air mattress type pump works great, as shown in Figure **G**. These pumps deliver a large volume of air at a fairly low pressure, and this is precisely what we need for this application. A bicycle pump may also be used, though it will not deliver the large volume of air a mattress pump will. Plug the tip of the pump's hose into the fitting in the side of the horn's base frame, and secure it in place with more tape, creating an air-tight seal. Now, by simply actuating the pump, you can annoy your neighbors with your new, loud, two-tone air horn. ◗

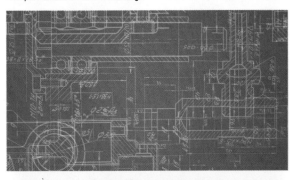

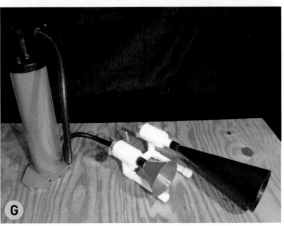

G Completed dual air horn assembly with air mattress pump

Tracking Further:
Using Ingenuity – Self-Tapping PVC with a Metal Pipe Fitting

As a maker, there will be instances in a project where you need to use creativity or ingenuity to solve a specific problem. Sometimes, you can easily modify a purchased item to adapt it such that it will work for your specific project.

The following example of needing to be able to screw a pneumatic fitting into untapped PVC is a real-world example of a roadblock that Brian faced during a specific project. Brian was working on a pneumatic cannon, nearly finished, when he realized that he had not installed the nifty little air fitting purchased to get compressed air into the cannon. Oops!

Rummaging through his various taps and drill bits, he quickly realized that he did not have the correct NPT tapered pipe tap required to mount the fitting. So, Brian was stuck and very frustrated that one little fitting would prevent the test firing of the otherwise completed cannon.

Contemplating his options (and not wanting to interrupt his project momentum by taking a trip to the hardware store to get the proper pipe tap), he thought about what a "tap" actually is. Realizing that it is simply a very hard, threaded screw-like tool that is used to cut threads into a softer material, he formulated the following method for using what he had to tap the hole.

Referring back to the pipe coupler we found for the air horn project that was missing an input port, similarly, we need to create a hole to mount a pneumatic fitting directly into the PVC part. In other words, we need to "tap" a hole for the fitting to be mounted. "Tapping" is simply cutting screw-threads into a material. If you had the "optimal" tools for tapping, you would simply grab your national pipe thread taper (NPT) pipe thread tap tool and its associated tap drill, and drill and tap the hole. Figure **A** shows a typical NPT pipe tap and its associated tap drill bit.

Most people don't have these tools lying around, as they are very specific and can be quite expensive. However, don't fret! There is another way! (And, unless you are planning on tapping a bunch of holes with the same pipe

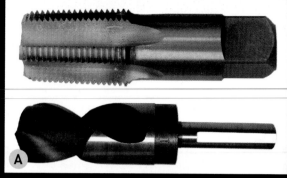

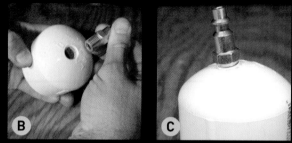

Ø59/64" tap drill bit with ¾"-14 NPT tap

Using a metal air fitting to self-tap a hole in PVC

NPT (National Pipe Tapered-Thread) Tap Drill Size

NPT Size	Tap Drill Size (in.)	Decimal (in.)
⅛ - 27	²¹/₆₄	0.328
¼ - 18	⁷/₁₆	0.438
⅜ - 18	³⁷/₆₄	0.578
½ - 14	²³/₃₂	0.719
¾ - 14	⁵⁹/₆₄	0.922

OFF

Using calipers to measure thread root diameter

D

thread, don't go out and buy a pipe tap for your PVC project.) Instead, we can use an inexpensive, readily available metal pipe fitting for the job. These pipe fittings can be found at your local hardware or DIY store.

Don't get a plastic fitting. The fact that these are metal allows the fitting to be used as a tap tool! This works because the very hard metal threads are able to cut threads into the much softer PVC.

Figures **B** and **C** shows a 2" PVC pipe cap fitting tapped using a metal ¼-18 NPT pneumatic fitting.

To start tapping the PVC fitting, figure out what pilot hole (technically known as a tap hole) is needed to accommodate the fitting you will be using. This tap hole (slightly smaller than the threads of the fitting) will ultimately serve as the base for the threads.

Once you know your fitting thread size (in this case, NPT size), you can utilize standard reference tables, like the "NPT Tap Drill Size" table, to figure out the drill bit size to use as the tap drill (reference tables like this can be found in hardware stores, in engineering books, or on the internet). For example, the fitting thread size used in the figure above was labeled ¼-18, so the drill bit that we needed for the tap hole (according to the NPT table) was ⁷⁄₁₆" in diameter.

Sometimes, you don't have a table like this one available when you are in the middle of a project and need to tap a quick hole. Using calipers, simply measure the root diameter (the innermost diameter of the threads or the trough of the threads) of the fitting. That said, we need to give you a word of caution here. In the case of NPT pipe fittings, the threads are tapered. This means that the root diameter will increase as you move up the fitting, so measure the root diameter about a ¼ of the way up the threads from the end of the fitting (not from the head), as shown in Figure **D**. You will want to use the closest size drill bit you have to the root diameter measurement, and then drill a tap hole in the PVC where you want your fitting to be located.

Confession time: When tapping a soft material such as plastic or wood, we (as makers) typically do not even measure the thread root diameter to figure out the tap drill needed. With the metal pipe fitting in hand, we

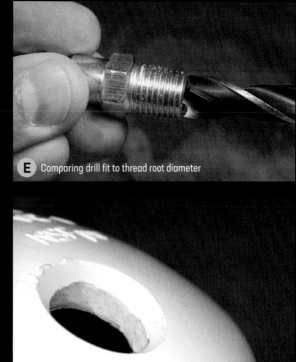

E Comparing drill fit to thread root diameter

F Chamfered tap drill hole

visually match the drill bit that is closest in diameter to the root diameter of the fitting threads, as shown in Figure **E**. In this way, we determine the drill bit we use as the tap drill, and this method seems to work just fine.

It is helpful to chamfer the tap hole slightly using a countersink bit, rotary tool or even just sandpaper. To chamfer a hole means to make the edge of the hole angled or slightly tapered. This gives you a nice angled lead into the hole to be tapped. This is shown in Figure **F**.

Once your tap hole is drilled and chamfered, align the metal fitting (which you are using as your tapping tool) with the hole. Then, using a wrench, start turning the fitting clockwise while simultaneously applying axial force (i.e., straight toward the hole) to cut threads into the PVC. In simplerterms: Push the fitting into the tap drill hole while rotating it in the proper direction for the threads you are cutting. The metal threads will dig into the PVC and cut (screw) their way into the part. Continue to screw the fitting into the PVC until it gets moderately hard to turn. At this point, unscrew the metal fitting and you will see that the tap drill hole in the PVC is now threaded.

G Metal pneumatic fitting used to self-tap wood

Apply a few wraps of Teflon tape to the fitting and screw it back in the hole for an airtight fit. Great job! Your tapping is now complete.

By the way, you can also use this method for cutting pipe threads in wood (Figure **G**). In fact, this method works well for tapping many soft materials like wood or plastic, considering that the metal fitting being used as a tap is much harder than the material being tapped. But, purchased taps are made out of hardened carbon steel, and are necessary for tapping metals, such as aluminum and steel.

Going back to Brian's real-world air cannon project that we mentioned above, his primary reason for using this method was to prevent interruption of his project momentum. Going to and from the hardware store to get the proper pipe tap would have taken a good hour or more, and this certainly would have killed his project momentum.

Makers get to a particular point in a project where we need and want to continue working on it, especially through a critical or exciting part. Needless to say, the point at which Brian was about to be able to test the pneumatic cannon was one of great momentum and excitement. So he very quickly rejected the idea of stopping what he was doing and going to the hardware store to get a tap. Then he formulated the idea of using the actual fitting needed as a tap to make the threads in the PVC cannon. It worked like a charm!

This is an example of where creative, maker ingenuity can be used to solve project-related problems. Brian has made many tapped holes using this method since he originally figured out this little trick.

Modularization of Design

Large, complex projects, such as the Tot-Size Tank, are exciting to contemplate, but can quickly become overwhelming. The idea of trying to design and build an entire tank is a daunting prospect - your authors speak from direct experience. So, how do you eat an elephant (or, how do you break a huge project down into manageable chunks)? The answer is, as you might imagine, one bite at a time.

"Modularization of design" is simply the idea of breaking down complex, large projects into smaller, more manageable pieces. When initially thinking about the tank project, Brian quickly realized that he would have to break it down into manageable pieces or modules. For the purpose of this book, a module is defined as a subassembly of parts that, when put together, forms a complete system that can be built and tested by itself. Each module is a piece of the overall larger project. In other words, the large, complex project is built or comprised of a series of modules. The Tot-Size Tank is comprised of three main mechanical modules: Module 1: Tracks, Module 2: Frame and Drivetrain, and Module 3: Body (Figure **A**).

In designing and building the tank, Brian started out with the tracks (Module 1 - Figure **B**). His goal was to build the tracks so that each one could be built and tested independently without requiring any other module, or part of the tank, to be completed or even designed. The challenge of the track module was how to support, tension and drive the conveyor chain. Each track is built around a central backbone, on which all of the other components are supported. Bolted to the backbone at both the front and back end are dual bearing blocks that serve to support the drive axle at the front and the idler axle at the rear, and attached to each axle is a sprocket. The sprockets transmit power to the track, support the track, and allow for the tensioning of the track. Pulleys attached to the sprockets transmit drive torque to the drive axle, support and guide the track, and supplyy braking force through the idler axle. Lastly, the riser blocks and lower track support/guide support the tracks at the ground, and provide for a better geometric path through which the tracks move.

Next, Brian focused on supporting and driving the track

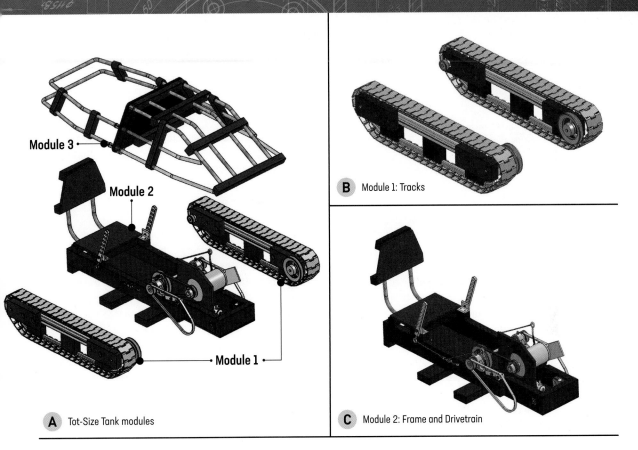

A Tot-Size Tank modules

B Module 1: Tracks

C Module 2: Frame and Drivetrain

system, hence Module 2: Frame and Drivetrain of the Tot-Size Tank (Figure **C**). The tracks needed a way to be supported and driven. Brian knew that he needed a structure that could simultaneously support the tracks at the proper width, provide for a way to house and mount a drivetrain, have a means to mount both driver controls and eventually, a body, and it had to accommodate a kid-size driver. Wow! That was a ton to think about for one simple frame, and it would have been overwhelming to try to consider all of those constraints at once. He broke this large concept down, mentally, into smaller, manageable chunks or sub-modules.

First, he laid out some of the major components that had to be attached to this frame or integrated with it. Mainly, this consisted of the motor and one completed track, as he had only built one track assembly at this point in the project. After placing the motor and track down on the workbench, he was able to roughly visualize what the finished tank would look like, and how the drive system could be laid out. He sat his son on the bench toward

the back of the track, as if he were driving the future tiny tank. This illustrated roughly how wide the tank should be. Armed with this very rough, initial idea of how the geometry of the frame should be laid out, Brian was able to begin designing Module 2.

This exercise, of physically laying components of a project out in their rough relative position (looking at an aspect of a more complex project in a hands-on, 3D sense), can really help bring together a potentially complicated design.

Module 2 turned out to be a nice balance of reasonable weight, good stiffness, and relative simplicity. At the core of Module 2 is the frame. Attached to the rear portion of the frame are a seat and controls for the driver. The motor, pulleys, shafts, and belts that make up the drivetrain are mounted in the frame from the middle to the front of the frame, and there are provisions to mount the body, Module 3.

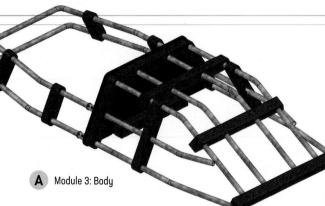

A Module 3: Body

Object-Oriented Programming:
Another Illustration of Modularization of Design

No matter how complex your project is, you can break it down into small, bite-sized chunks that ultimately fit together as the final project. The idea is to make as much of your project as modular as possible, so that each subsystem is an independent component that you can use and re-use whenever you need to. In software development, this is called object-oriented programming, or OOP.

OOP allows the programmer to develop tools and functions that can be applied to not just one, but multiple applications and projects. In some cases, making an upgrade to a component can have the added benefit of upgrading all the applications that use the same component. By the same token, in treating each component as its own individual "black box," OOP dictates that any subsystem is self-contained enough to have known inputs and known outputs. This means that if we want to swap out the subsystem (for example, our tank's track systems) for an improved version, we can, without heavily impacting the other subsystems.

Module 3: Body is the final module of the Tot-Size Tank (Figure **A**). There were three primary constraints that had to be met by the body. Number 1: It had to keep the driver and others around the tank safe from the moving drive components. Number 2: It had to be rigid, and keep its shape when mounted to the tank. Number 3: It had to be aesthetically pleasing – it needed to look good.

Module 3 also had a few secondary constraints (secondary aspects of a design that are not absolutely necessary but would really be nice to have). These constraints were that it needed to be fairly lightweight, and be made of easy-to-get and low-cost materials. In the course of conceptualizing and building the body, both of these secondary constraints were achieved. The body's framework is built from inexpensive, lightweight electrical conduit supported by lengths of common 2"×2" lumber.

Throughout this book, you will learn more details about the Tot-Size Tank, such as what makes up each module and why particular materials and items were chosen. However, this breakdown of the three Tot-Size Tank modules is intended to illustrate the importance of dividing a complex project into bite-sized pieces.

Modularization of design helps you design and build large projects without getting overwhelmed. It also helps you build and test subsystems of your design as you go along. In this way, you can work the kinks out of a more manageable section of the project. Hopefully, when you assemble all of the modules into your complete project, you will have far fewer issues as a whole.

Adopting the OOP methodology allows programmers to build components for other potential future projects. In OOP this is the development of a "library" of tools and functions that are common to all applications. OOP speaks to not just modularity, but also reuse and the concept of "inheritance." In a lot of ways, as makers, reuse and inheritance are closely related. If we design a propulsion system, can we make two of them and use the same system twice? (This is an example of both concepts.) If we take our project apart, can the same propulsion system be transplanted into another project with minimal modifications (i.e., reuse)? If our propulsion system has commonality with all others derived from the original, we have a loose interpretation of inheritance.

LET'S LOOK AT SOME EXAMPLES

- A wheel and its bearing is a subsystem. Can you make the bearing and mount for the wheel something that you can use in more than one project, or in or in more than one place in your project? On the Adult-Size Tank, the same set of bearing blocks are used for both the main wheels and the tensioning system that keeps the track taut. The same wheels are used as the "road wheels" as well as the drive wheel and the idler, with the only difference being the number of teeth on each wheel. Otherwise, the wheels are exactly the same design that Brian originally modeled.

- In the Tot-Size Tank, the system that enables the vehicle to turn also permits it to go forward. In the Adult-Size Tank, the steering system uses the same electric motors, switches, and components to operate each side. Not only are the motors able to drive the tank forward, but running each motor independently eliminates the need for a steering system. A throttle on each motor controls steering by changing the speed at which each motor turns.

- In one Go-Kart design, Samer used the same bearing concept on the drive (rear wheels) of the Go-Kart as he did on the bearing that holds the steering wheel.

- On Version 2.0 of the Adult-Size Tank, instead of using multiple materials for the frame, Samer used aluminum T-slot extrusion for the entire frame, adjusting the dimensions and positions of various frame components by simply tightening and loosening screws. Could you do the same, perhaps with lengths of perforated angle?

- Samer has used the same drive system combination in several Go-Karts, an earlier version of the Adult-Sized Tank, and even a half-track. Granted that this pretty much fixed the width of each design to 24", but this was an engineering compromise worth making. In fact, as each project was built, the drive system, wheels, and motor were harvested from the previous version of the project.

In this chapter, we wanted to introduce you to the process of how we, as mechanical engineers and makers, design and bring a project to reality. The project realization process we introduced is a general suggestion of the process steps that we take. These steps are not linear: They should be interchanged and revisited as often as needed. Sketching, prototyping, and testing help to refine project specifications, and the fabrication methods need to be considered in the design process.

With these process steps in mind (and some practice under your belt from the Air Horn project), we can now dive into the details of making, starting with what to make things out of. Let's move on to materials!

Jeffrey Braverman, Heo Syadia

3

Materials Selection – Plastics, Woods, and Metals... OH MY!

Materials — the elements, constituents, or substances of which a thing is or can be made —have been important to humans since ancient times. They even define historical eras: Stone Age, Bronze Age, Iron Age, Roman (lead) Age and the Industrial (steel) Age. Humans utilize all types of materials, including stone, cement/concrete, wood, ceramics, fiber/cloth, glass, metal, paper, plastics, rubber, magnetic materials, and biological materials.

Although all materials are important and useful for certain situations, some materials are more commonly used by mechanical engineers and makers than others. These include wood, metals, glass, plastics, foam, and some composite materials. As makers, you will want to be familiar with the properties of these different materials, so that you can make the best materials selections possible for your projects.

Material Properties to Consider

Choosing the right materials when designing and building a new project is vital to its.success. In order .to avert avoidable problems down the road, you should take the time to consider the following material requirements:

- **Function:** Will the material provide the function that is needed?
- **Strength & Durability:** Will it hold up to the stresses and forces that it will be exposed to over time, without deforming or failing?
- **Workability:** Can the material be easily modified, molded, or fabricated using the tools you have on hand, or are willing to buy or borrow?
- **Aesthetics:** Does the material add to the appearance of the final project, without negatively impacting other aspects such as the durability or cost?
- **Safety:** Is this material safe to work with using the tools and working conditions that you have available?
- **Cost:** Is the material cost-effective, without negatively impacting other aspects such as the function, safety, and aesthetics?

Each material has a unique combination of these properties, and is evaluated in the discussion of raw materials below. Selecting materials that provide a proper balance between all of these aspects will directly impact the success of your projects.

A Bowl made from hardwoods

B Softwood lumber (pine)

C Plywood

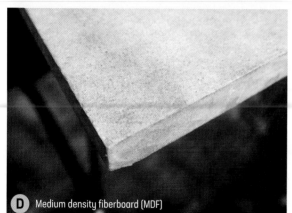

D Medium density fiberboard (MDF)

Types of Materials

There are many materials available to makers, and each material's properties recommend it to certain situations over others. Here, we survey the materials you are likely to come across as a maker, and examine some of their unique properties and common uses.

WOOD

Wood and wood-based products are among the most common materials for makers. Wood comes in many formats, such as blocks, boards, dowels, and sheets. It is incredibly versatile and easy to work with. It can be used for prototyping, as well as for finished products. There's a whole industry of tools and learning materials based on woodworking.

Common types of wood materials include:

- **Hardwood:** Hardwoods (Figure **A**), like oak and hickory, are dense, heavy, and sturdy. Their tightly-packed growth rings produce compact, distinctive grains that are great for withstanding heavy wear, and can also be quite beautiful. Hardwoods are heavy, solid, and rigid. This type of wood tends to not be as flexible as softwood. Therefore, it tends to break rather than bend when overloaded. These woods are usually more expensive and less plentiful than softwoods. Typically, hardwoods are used for furniture, cabinetry, hardwood flooring, and other areas that require beauty, strength, or durability.

- **Softwood:** Softwoods (Figure **B**), such as pine and cedar, are more pliable and flexible than hardwoods. They are also lighter and less dense. But they are still very strong structural materials. These woods are relatively inexpensive, plentiful, and easy to work with. Softwoods are commonly used for framing buildings, moldings, shelving, and home woodworking projects. Most of the time, softwoods to be used for outdoor projects will need to be "pressure-treated." Pressure treating is a chemical process that makes the wood resistant to rot, insects, and mold. Treated wood is more expensive than non-treated wood.

- **Plywood:** Plywood (Figure **C**) is a laminate of thin layers of different woods with crossed grain, meaning that the lines of weakness (grain) for one laminate layer

Staying on Track:
Distinguishing between Hard- and Softwoods

Biology 101: Softwoods are coniferous (i.e., evergreen) trees, and hardwoods are deciduous (broadleaf) trees. But, how does that help you figure out which are which at the store? It is difficult to tell if a block of wood is a hard- or softwood just by looking. Fortunately, there are a few ways you can tell the difference.

- Softwoods will typically have a wider end grain pattern than hardwoods, because softwood trees grow much faster. So, one way to tell the difference between softwood and hardwood is to look at how dense the end grain pattern is. Hardwood end-grain patterns will be considerably finer than that of softwoods. The figure below shows an end-grain pattern example for hardwood, on the left, and softwood, on the right.

- When you run your fingernail along the surface of the wood and it makes a dent in the wood, then it is most likely softwood. Hardwoods are typically harder than your fingernail and will therefore not be easily marked by it.

- Hardwoods are typically not as aromatic as softwoods. So, if you smell something that reminds you of the holidays (i.e., pine or cedar), then it's definitely a softwood!

End-grain pattern example, hardwood (left), softwood (right)

are 90° to the grain of the layers above and below. This cross-grain laminate effect makes plywood relatively lightweight and very strong. Plywood sheets can be bent and shaped, and some offer an outer layer that can be sanded for a nice, finished look. It is used in house construction, furniture making, and even aviation, where it is used to cover the frames and flying surfaces of light aircraft.

- **MDF:** Medium Density Fiberboard (MDF) is a synthetic composite (Figure **D**) made with wood fibers and resin. MDF is very sturdy and takes paint quite well, so it is commonly used for shelves and cabinet doors. The downsides to MDF are that it is very heavy and does not play well with water - it will deteriorate (i.e., swell and come apart) when it gets wet. MDF can be held together with glue and screws, but the screws will easily strip out under heavy loads.

A Aluminum stock

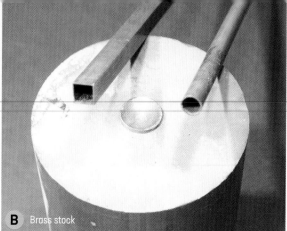

B Brass stock

C Copper tube

D Mild carbon steel

METALS

Metals are another common material that makers can use in their projects. Metals can be used in pure forms or as alloys. An **alloy** is a mixture of metal elements, combined to make the resulting material stronger and/or more resistant to corrosion.

Metals are generally classified into two categories, **ferrous** and **non-ferrous**. Not to get too deep in materials science, but the primary difference between the two is simply that ferrous metals contain iron (Fe), where non-ferrous metals do not. Here are just a few types of metals available, representing each category:

NON-FERROUS METALS AND ALLOYS (WITHOUT IRON)

- **Aluminum:** a strong, lightweight, corrosion-resistant, and easy to cut, machine, and form. It comes in a variety of thicknesses, shapes, and strengths, and can be purchased from a variety of vendors (Figure **A**). Sheet aluminum is great for skins and coverings, such as on airplanes or enclosed trailers. Plate aluminum is more structural and can be cut, bent, and welded.

- **Brass:** (Figure **B**): an alloy of copper and zinc, and has been used as an engineering metal for centuries. Brass can be bought as sheets, tubes, and bars, and is very **malleable** or ductile, meaning it can be formed to a specific shape without breaking. It is also rather soft, so is easily turned/cut on a lathe. It is used to make home decorations (like candlesticks), firearm cartridges, and musical instruments such as trumpets and tubas. Its thermal properties lend themselves to applications like hobby steam engines, and its anti-corrosion properties are good for marine applications. Compressed air fittings, spacers, and other hardware are also commonly made of brass.

- **Copper:** (Figure **C**): an excellent conductor of electricity, and is malleable like brass. Copper is commonly used in water pipes and electrical wiring (though not together - that would be dangerous!). Copper is easily soldered or brazed with little more than a blow torch and some flux and solder. (**Brazing** is a metal-joining process where metal of a lower melting point is heated to melting, and allowed to flow into the joint between the metal pieces being bonded.) Copper oxidizes very easily, developing a blue-green patina

- the Statue of Liberty is skinned in copper sheeting, which is why she has a bluish-green hue.

FERROUS METALS AND ALLOYS (WITH IRON)

- **Steel:** an alloy of iron, carbon, and other metals. Steel is very hard and will typically corrode (rust). Steel is used extensively as a structural material. It can be a bit more difficult than other metals to machine, but it is workable with the correct tools. There are many variations of steel, each with unique properties:

  - **Mild Carbon Steels:** iron-carbon alloys containing between 0.05–0.25% carbon. Mild carbon steel is the most commonly-used steel because it is cheap, strong, malleable, magnetic, and not brittle. Its drawback is that it is prone to rust (Figure **D**).

  - **Stainless Steels** are iron-based alloys containing a minimum of around 10.5% chromium (Figure **E**). The chromium allows the steel to to resist corrosion. When compared to mild carbon steel, stainless steel is, in general, not as strong as mild carbon steel, but it tends to be harder and requires less maintenance. It has a more attractive appearance, but it is also more expensive.

  - **Chromoly** (Figure **F**): an iron-based steel that also contains chromium and molybdenum. The density of chromoly is equivalent to that of mild carbon steel, but it has nearly double the strength. In other words it has a much better **strength-to-weight ratio** (i.e., specific strength). In a practical sense, if you are making a frame out of tubular metal, you can use a much thinner tube if using chromoly than if you were using tubing made from another steel alloy. It has high tensile strength, but is more brittle and easier to damage than other steel, and does not have the corrosion resistance of stainless steel. Chromoly is widely used in the bicycle and aircraft industries.

  - **Cast Iron** (Figure **G**): a brittle, non-malleable iron-carbon alloy that melts at relatively low temperatures, and is easily cast in molds. Finished cast-iron objects are very hard and cannot be bent, but are easily machined, resistant to wear, and are relatively inexpensive. Items such as cookware, pipes, and some automotive parts are made from cast iron.

E Stainless steel

F Chromoly tube

G Cast iron cauldron

A Broken annealed (plate) glass

B Broken tempered-glass window

C Safety glass (car window)

Borosilicate (Pyrex) glass beaker

D

GLASS — Glass is a non-crystalline, amorphous solid (between a liquid and solid state) that is usually made of silicate, but can be composed of other materials. It comes in an amazing variety of forms, and some of it can be quite tough. Types of glass include:

- **Annealed:** Borrowing a common term in metal fabrication, annealed glass is glass that has been allowed to cool slowly after it has been formed into its final shape. Cooling slowly allows internal stresses to be relieved, making the resulting object more durable. Annealed glass can be very dangerous when it breaks, because it shatters into many sharp-edged shards and slivers. Glass used in household objects, like the picture frame glass in Figure **A** (sometimes called "plate" glass), bottles, and inexpensive drinkware are examples of annealed glass.

- **Tempered:** Tempered glass is processed (either chemically or with heat) so that the outer layers of the glass are in compression while the inner layers are in tension. This gives the glass greater strength and allows it to crumble into small parts on impact, instead of shattering into very sharp and dangerous shards. Examples of tempered glass include shower doors, side and rear car windows, and doors inside stores and office buildings (Figure **B**).

- **Laminated:** Laminated glass is made up of multiple layers of (usually tempered) glass, with a plastic "interlayer" between each glass layer. The interlayer bonds to the glass layers, and this bonding prevents the shattered pieces from exploding out from the interlayer in the event of an impact. Because the shattered pieces are restrained by their bonds with the interlayers, and are not likely to explode out from the impact, laminated glass is also called "safety glass." Car windshields are made of laminated tempered glass, which becomes evident when they are hit by a rock or other hard object. A crack spreads outward from the impact area, producing a "spider web" pattern on the windshield as the glass bits remain adhered to the interlayer (Figure **C**).

- **Borosilicate:** Borosilicate glass is made using silica and boron trioxide. The most important property of this glass is its resistance to thermal shock, which is why it is used in glass cookware, laboratory glassware (Figure **D**), 3D printer surfaces, and the glass mirror elements of reflecting telescopes.

Any sheet glass, whether a window pane or the table of a scanner, is glass you can use for something. Borosilicate glass is useful as a build surface on 3D printers, and other sheet glass can be used as windows and viewports in enclosures. Glass can be harvested from many old components, such as old house windows and doors, all-in-one printers, and picture frames.

Glass bed on 3D printer

E Polyester resin necklace pendant

F PVC

PLASTICS

PLASTICS — Plastics (or polymers) are available as rigid or flexible materials. They are classified into two basic, broad categories: thermosets and thermoplastics

THERMOSET PLASTICS

Thermoset plastics are initially formed and then "set" with heat. Once initially formed into a given shape and set, a thermoset plastic cannot be reformed through the application of heat. Phenolic, silicone, and epoxies are examples of thermosets. Commonly used thermoset plastics include resins, which are two-part viscous fluids (a base resin and a hardener) that, when combined, generate a thermsetting reaction that hardens the plastic into a very tough material.

- **Polyester Resins:** Polyester resins are easy to work with and relatively fast curing. They are frequently used in fiberglassing, and are also used in craft and jewelry making (Figure **E**) since they are inexpensive and commonly available. Another very common use of polyester resin is in laser printer toner, where the resin is mixed with carbon and a dye, and then melted onto paper.

- **Epoxy Resins:** Epoxy resins are up to four times stronger than polyester resins, and are more resistant to physical and chemical wear as well as environmental damage. They are also more water-resistant than polyester resins. This increased strength and resistance makes them more expensive, but they cure faster and do not have the unpleasant smell of their polyester relatives. Because of their excellent strength and waterproofing properties, epoxy resins are commonly used in many fiber composite materials, which are discussed later in this chapter.

THERMOPLASTICS

Thermoplastics are plastics that can be re-melted and molded into something new. Filament used in 3D printers is a good example of thermoplastic. Nylon, PVC, polycarbonate, polyethylene, polypropylene, and acrylic are all thermoplastics. This class of plastic can be bent or formed into different shapes through the application of heat. Once cooled, the plastic will retain its shape.

- **Polyvinyl chloride (PVC):** comes in two basic forms: rigid and flexible. PVC water pipes are an example of the rigid form (Figure **F**). Wire insulation is commonly made from the flexible form. PVC is a great maker

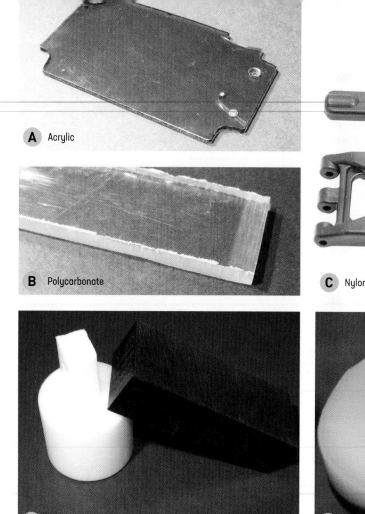

A Acrylic

B Polycarbonate

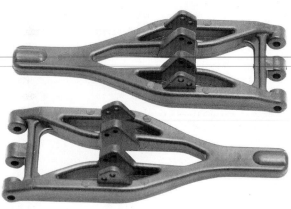

C Nylon RC car suspension arms

D Polyoxymethylene or Delrin

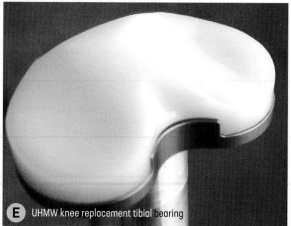

E UHMW knee replacement tibial bearing

material because it is fairly strong, readily available, and relatively inexpensive. PVC is also quite easy to cut, drill, sand, and (with the application of a little heat) bend. As you will see throughout this book, PVC can be used to build a wide variety of projects.

- **Acrylic (Plexiglass):** This is a great material to use for applications where translucency or transparency is important. Sheets acrylic (Figure **A**) can be purchased in various thicknesses and sizes, and makes an excellent material for transparent panels. It is also relatively easy to work with, though it does have a tendency to be brittle. You have to be a bit careful when drilling or cutting acrylic, as it will crack easily. With the application of heat, acrylic can become quite malleable and can be bent to the desired shape or accept embedded components, such as thermoplastic inserts.

- **Polycarbonate (Lexan):** This is really tough stuff. Bulletproof windows are typically made out of polycarbonate, as it is both very tough and transparent. It is a very strong material that is easy to cut, drill, and tap to screw fasteners directly into it. Unlike acrylic, it is not brittle and will not crack easily. Polycarbonate (Figure **B**) is typically more expensive than acrylic, and it cannot be easily shaped by heating it up. It can be bent by "cold breaking," meaning that it can be bent at room temperature by over-bending it past whatever angle you wish to achieve, as it will spring back somewhat, but not all the way. Tinted polycarbonate is sometimes used to block wavelengths of light. For example, green-tinted polycarbonate can block harmful blue light, which is why it is excellent for laser engraver enclosures and safety glasses. Other examples of commonly-used items made from polycarbonate are water bottles, CD/DVDs and plastic drinkware.

- **Nylon** is a family of plastics available in a variety of forms, such as sheet, blocks, rods, bars, thread, and filament. It and has a wide range of uses, from synthetic fabric to structural applications (Figure **C**). Nylon filament is also available for fabricating objects with 3D printers.

- **Delrin:** Also known as acetal, polyacetal, and "Delrin," polyoxymethylene (Figure **D**) is a tough thermoplastic that is suitable for precision parts. Polyoxymethylene is widely used in the automotive industry as well as for ball ball bearings, small gear wheels, fasteners, eyeglass frames, and ski bindings. Polyoxymethylene is commonly an opaque white, but also can be found in a variety of colors.

- **Ultra-high Molecular Weight Polyethylene (UHMW):** This type of plastic is really cool stuff. It is an opaque plastic that has excellent wear properties and a low coefficient of friction, making it a great choice for bearing applications. In a total knee replacement, UHMW is used as a bearing material within the knee joint (known as a tibial bearing, as shown in Figure **E**). UHMW can be a bit tricky to work with, as it is somewhat soft and gummy to cut and drill.

- **3D Printing Filament:** 3D printing has become common and easily accessible to makers, both for prototyping and for finished products. TWhile there are many kinds of 3D filament, the most common filaments are:
  - **Polylactic acid (PLA):** has a low melting temperature and doesn't require a heated bed (although it does help to have a bed heated to 50-60° C). It can be made from cornstarch or sugarcane and is, therefore, biodegradable. PLA is relatively more dense (heavy) than other plastic filaments, and does not easily warp during printing. The latest formulations of PLA rival ABS in strength.
  - **Acrylonitrile butadiene styrene (ABS):** melts at a high temperature and requires a heated printing bed. It is generally less dense (lighter) than PLA, and frequently needs effective cooling and bed treatment to keep the initial layers from warping off the print-bed, especially for larger surfaces. ABS can be more difficult than PLA to print with, but the finished product will be more durable.
  - **Exotics:** Today, filaments can be wood (Figure **F**), metal and even aramid/Kevlar-infused. Imagine

F Nefertiti bust 3D printed with a wood-infused filament

being able to fabricate something that can inherently conduct electricity (like a printed circuit board), or a component that can be magnetized without having to install a separate magnet. Filaments infused with Kevlar or carbon fibers add strength as a property without having to use additional layers of cloth, vacuum-forming, or other fabrication techniques.

One thing to keep in mind is that 3D printed objects can be treated with resins that are self-leveling and fill in any imperfections on the surface of a print. These resins also provide added strength, durability, and a smooth finish, with limited need for sanding, if any, sanding and other post-print processing.

A Fiberglass pool insert

B Carbon fiber race car rear wing

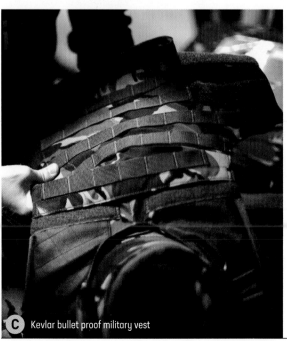

C Kevlar bullet proof military vest

FIBER-REINFORCED COMPOSITES

Fiber-reinforced composites are materials made of a mixture of fibers (glass, aramid/Kevlar, or carbon) and a matrix material (typically, a thermoset plastic resin), combined to increase strength and durability. The combination of the orientation of the fibers and the resins used to bond the fibers impact the overall properties of the material. In general, fiber-reinforced composites have a high strength-to-weight ratio, offer good vibrational damping, are typically waterproof, and are resistant to fatigue, extreme temperatures, corrosion, and wear.

- **Glass fibers (Fiberglass):** Most fiber-reinforced composites are made from glass fibers. Glass fibers are manufactured as strands, and can be twisted together into yarn or formed into cloth. Glass fiber weighs more and is more flexible than carbon fiber (the next most commonly used fiber). This flexibility allows glass fibers to be more resistant to impact and damage before breaking. Glass fiber composites are used in a variety of applications, including the .marine, airplane, and automotive industries, and even home swimming pools (Figure **A**).

- **Carbon fibers:** Carbon fibers are also commonly used in fiber-reinforced composites, but are more costly. They are manufactured as continuous fibers that can be formed into tape or cloth, and are widely used in high performance applications such as race vehicles, aircraft, and spacecraft. Carbon fibers provide high strength and stiffness, and are resistant to thermal expansion. They are stronger than glass and aramid fibers (described below), but are also less resistant to impact before breaking (Figure **B**).

- **Aramid fibers:** Typically known as Kevlar, aramid-fiber composites are commonly used for military armor and personnel protection applications (Figure **C**). Aramid fiber has excellent impact resistance and relatively good flexibility, making it an ideal material for protection.

FOAM

Foam is an especially useful material for makers, as foam is lightweight, provides good core structure, and is relatively inexpensive. Foam is easy to work into complex shapes that would otherwise be very difficult to fabricate with metal or wood. Structures made with a foam core and fiber-reinforcing composite "skin" rival metal from a strength-to-weight and **stiffness-to-weight** (i.e., specific modulus) standpoint. Light aircraft wings, which need to be very lightweight, strong, and durable, are commonly made this way to reduce weight without sacrificing strength.

Makers typically use rigid foam, although there are certainly other types of foam such as the soft, squishy foam used in seat cushions or pillows. The main types of rigid foam that most makers will find useful (and at times indispensable) for their projects are white foam, blue or pink foam, foam board, and two-part foam.

- **White:** White foam is made up of a very large number of small (1.5mm to 4mm) polystyrene beads that have been heated and compressed into large sheets. It is mostly air by volume, so it is lightweight but not rigid. White foam is a great material to use when weight is an important consideration, such as when building a model airplane. Figure **D** shows a small, very light glider that is made completely of white foam. This type of foam can also be used as a core material that can be covered or skinned by another material, such as wood or fiberglass, to make a very rigid, light structure. Some heavier remote-control planes use foam as a core in their wing construction. White foam is the least expensive of the foam types that most makers encounter, but white foam can be quite difficult to shape by sanding or carving. Due to the granular, bead structure that makes up this kind of foam, it tends to crumble when you try to sand or carve it. However, this type of foam can be fairly easily cut and shaped by a device known as a hot wire cutter. A **hot wire cutter** (Figure **E**) is simply a wire that is supported by an electrically-insulated bow and heated by passing an electrical current through it. The cutter cuts or shapes the foam by melting it in a very localized area around the wire as the wire passes through the foam. There is a video on YouTube from FliteTest that explains hot wire cutters and how to use them. The video link is youtube. com/watch?v=fi3CAtpvJJs.

- **Blue or Pink:** Blue or pink foams are basically the same material (different colors are made by different companies), but generally referred to as "blue foam." It is mostly used in construction as an insulation material. Blue foam is formed by an extrusion process that produces many strands of tightly-packed polystyrene

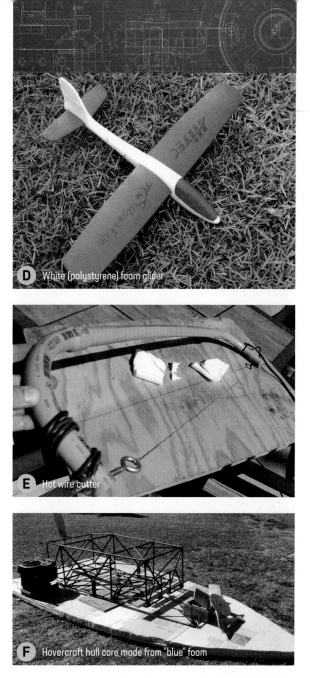

D   White (polystyrene) foam glider

E   Hot wire cutter

F   Hovercraft hull core made from "blue" foam

fibers, resulting in a much denser material than white foam. Considering the increased density, blue foam is significantly heavier, though much stiffer, than white foam. Blue foam can be cut via a hot wire cutter, but due to its makeup, it can also be shaped by other methods such as sanding and carving. For this reason, blue foam is a good choice of material to use where you need to form more complex shapes. Figure **F** shows the core of a hull for a hovercraft project, made of the pink version of this type of foam.

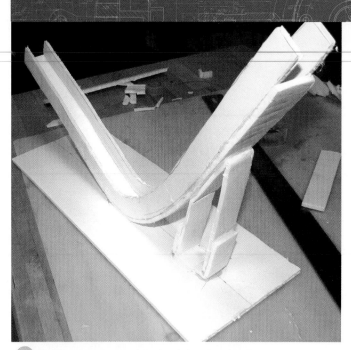

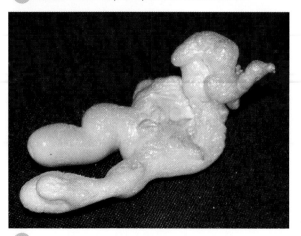

A Marble-coaster ramp and spiral made from foam board

B Dried and expanded two-part foam

● **Foam Board:** Foam board, sometimes referred to as foam core board, comes as thin sheets of foam covered on each side by heavy paper. It is a good example of a laminate material: a material made up of different layers fastened or glued together. Foam board is available in various thickness such as ⅛", ³⁄₁₆", and ½", with ³⁄₁₆" the most common thickness. Foam board is easily cut using a craft/hobby knife, razor blade, or even or even a hot-wire cutter. It is a useful material for prototyping and scale-modeling structures, such as architectural models. Figure A shows two parts of a marble coaster project made from foam board. A marble coaster is a series of tracks and other features that guide a marble from a starting point to a finish point. The structure on the left is a ramp. The structure on the right is a spiral feature.

● **Two-Part Foam** (Figure B)**:** Two-part foam is made of polyurethane. This type of foam starts as two liquids that, when mixed, react chemically and expand into a foam material. Two-part foams can be rigid or flexible depending on the formulation. Unlike the foam types above, two-part foams can be cast or molded to form specific shapes. These foams can also be shaped by carving, sanding, or cutting with a knife or hot wire cutter.

Staying on Track: Working With Materials

We frequently talk about being safe when working with tools. But we should also be very careful when working around certain materials. Most commonly, if you're working with materials that have sharp edges, like sheet metal, glass, or fiberglass, you probably know to wear gloves. And safety glasses are always essential when you're cutting materials. But here are some less obvious areas where safety precautions should be taken.

Pressure-treated Wood: This is a great, inexpensive, and durable material to work with, but the chemicals in it that make it resistant to rot and insects, also make it potentially harmful to you. Over the years, there have been several different types of pressure-treating chemicals, with some of the worst offenders (containing arsenic) banned from the marketplace. But, it is best to treat all pressure-treated wood the same way, especially if you aren't the original buyer, and don't know which chemical the wood is treated with.

- Wear a dust mask when cutting it to prevent inhalation of the chemically treated sawdust.
- Don't burn pressure-treated wood (especially as a bonfire or cooking fire), as burning releases chemicals into the air.
- Pull splinters out as soon as you are able, as they can irritate your skin. Better yet, wear gloves when working with it to avoid splinters altogether.

Plastics and Foams: These materials come in a very wide variety, so it is best to look up precautions based on the specific materials you are working with. In general, follow these guidelines.

- Don't burn plastics or foams, as the smoke is toxic, not to mention potentially harmful to the environment.
- If you are heating plastic to form it, work in a well-ventilated area. Wear a respirator as a precaution to prevent inhalation of fumes. These precautions should also be taken when using a hot wire cutter.
- Be aware that some plastics have a very rapid transition from soft to liquid when you attempt to heat-form them, meaning you might accidentally create a molten puddle under your work area. Take precautions to keep yourself protected in case this happens. For example, wearing sandals is probably a bad idea!

MDF: As previously described, medium-density fiberboard is an inexpensive and versatile material, but it is frequently made with a formaldehyde-based adhesive, so you should take care to protect yourself from it.

- One of MDF's most notorious attributes is the dust it makes when you cut it. This dust is extremely fine, and seems to go everywhere. Wear a dust mask every time you cut MDF, even outdoors.
- Don't burn MDF, because the chemicals in its adhesive can be toxic.

Fiberglass and Carbon Fiber: While these are wonderful materials, they have hazards of their own.

- Always wear gloves when handling them. The tiny fibers can easily turn into splinters.
- If you plan to cut fiber-based materials, wear a dust mask to protect yourself from breathing any fiber fragments that become airborne.
- Many of the epoxies used as a binding matrix for fiber-based materials, especially polyester resin used in fiberglass, emit toxic chemical vapors, so wear a respirator mask to protect yourself from inhaling them.

Old Paint: There was a time when common household paint contained lead. As the health hazards of lead became better understood, lead as a paint additive was banned in 1978. Reusing existing materials is a big part of making, so if you're recycling painted wood that dates from before that time, avoid cutting, scraping, or sanding it without a dust mask, or preferably a respirator. Don't burn it, either. If you're not sure, assume painted wood contains lead just to be safe. —*John Manly*

Staying on Track: Component & Repurposed Materials

Component materials are pre-fabricated items, such as bearings, gearboxes, pulleys, and pre-made brackets, that you may be able to incorporate into your project as-is. They can also be repurposed materials that may have already had other uses. Their original purpose may have been totally unrelated to what your project needs them for, but why not recycle and use what is already available?

You cannot imagine how many fasteners (screws, nuts, bolts, washers) and springs there are in an inkjet printer, not to mention motors galore! If you take apart an old printer, you will end up with little baggies of Torx drive screws, Phillips drive screws, countersunk head screws, pan head screws, self-tapping screws (Figure A)...the list goes on. Countless robots have come to life thanks to motors from toys the kids have destroyed, or old hard drives that gave up the ghost. What you use these parts for is up to you.

Sometimes, a recovered part is actually an entire system. For example, Samer has a habit of tearing apart old laptops and repurposing the LCDs as monitors for embedded systems (Figure B). He frequently finds himself using materials from other sources to make the enclosures, and the laptop's power brick ends up powering the same LCD that was in the laptop.

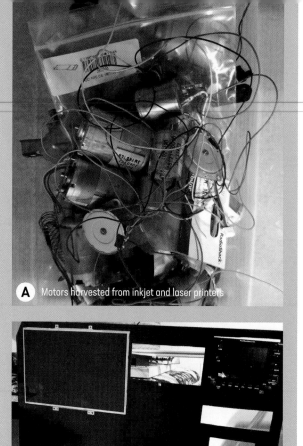

A Motors harvested from inkjet and laser printers

B Harvested laptop LCD panel for flight simulator

To see an example of disassembling and harvesting parts, specifically disassembling and harvesting parts from a laser printer, visit Instructables at instructables.com/id/Harvesting-parts-from-a-Laser-Printer.

Examples of Material Shapes
FOCUS ON METALS

Materials (including wood, metals, glass, and plastics) are available in many sizes and shapes (rod, bar stock, tubes, sheets, etc.), and can be used in a wide variety of applications. Here are examples of various shapes of metal materials:

- **Sheet:** Sheet metals are usually less than ¼" thick, and are sold in flat sheets or rolls. Thinner sheets can be rolled up, while thicker sheets cannot (Figure C).

- **Plate:** Plate metal, like sheet metal, comes in sheet form, but is ¼" thick or more (Figure D). Metal plate stock can be cut and used for brackets, gussets, and other structural elements.

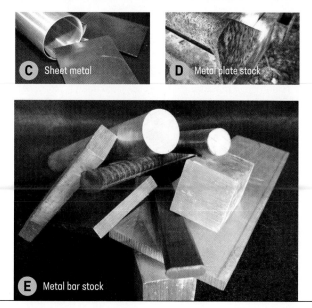

C Sheet metal

D Metal plate stock

E Metal bar stock

- **Bar Stock:** Bar stock is a fairly general shape category. Bars are solid lengths of metal available in many cross-sectional shapes such as round, square, rectangular, and hexagonal (Figure **E**). They come in a variety of sizes and lengths. Round bar stock is also called metal rod, and some rods are even threaded.

- **Tube Stock:** Tubes are hollow lengths of material that are available in many cross-sectional shapes, such as round, square, and rectangular (Figure **F**). They come in a variety of sizes and wall thicknesses. Round metal tubes may also be called pipe.

- **Aluminum T-slot Extrusion:** Extrusion is typically used to describe the process by which a material is made into a certain shape. But in terms of this discussion, extrusion will refer specifically to aluminum T-slot extrusion. Although there are many forms of aluminum extrusion, T-slot profile extrusion is the most popular (as shown in Figure **G**). Aluminum T-slot extrusion is extremely strong for its weight. The cross-sectional view of T-slot extrusion shows T-shaped tracks, which can be used with specialized T-slot nuts and bolts to attach things to them, or multiple pieces together.

F Tube stock

G Aluminum T-slot extrusion

Staying on Track: How Round Pipe and Tube Are Specified

Although both tubing and piping are round, hollow materials, they are intended for different purposes, and are therefore not specified the same way.

Round tubing is typically used in structural applications, so the outer diameter (OD) is the important dimension. This is why round tubing is specified by the OD and wall thickness, such as 1¼" OD × ⅛" wall thickness. To complicate things, wall thickness can also be specified as a "gauge", which can be looked up in a gauge table reference. So, the tubing listed in our example above can also be specified as 1¼" OD × 11 gauge, since 11 gauge is approximately ⅛".

Pipe is typically used as a transport vessel for gases and liquids, so the volume of the pipe is the important factor. This is why pipe is specified by the internal diameter (ID) and wall thickness. The specified ID is the average internal diameter within the tolerance band of the pipe, and is also referred to as the nominal pipe size, or NPS. (Although most makers seldom use large pipe, it is important to note that pipes larger than 12" in diameter have the NPS refer to the outer diameter.)

Wall thickness for pipe is called out as "schedule" dimensions, rather than as gauge. Most pipe found in a typical DIY store is schedule 40, which (for 1" PVC pipe) equals a wall thickness of 0.133". Sometimes, you can find schedule 80 pipe, which (for 1" PVC pipe) has a thicker wall of 0.179". Schedule 80 is intended for use in high-pressure situations.

Here's an example of how schedule can change the interior dimensions of pipe. A 1" ID-NPS schedule 40 PVC pipe has a theoretical ID of 1.049" while a 1" ID-NPS schedule 80 PVC pipe has a theoretical ID of 0.957". The volumes of these two pipes are slightly different, while the OD is the same for both (1.315"). The OD is the same so that the pipe (no matter the schedule) can fit within pipe fitting openings that are geometrically the same in size (even though schedule 80 fittings have thicker walls).

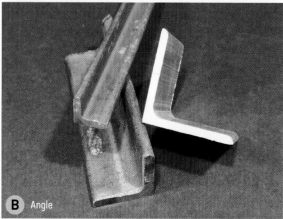

A C-channel

B Angle

- **Structural Shapes:** Structural shapes efficiently arrange material geometrically to resist forces, while keeping weight to a minimum. Different cross-sectional shapes are designed for different types of forces. Some shapes are designed to primarily resist bending, while others primarily resist twisting. Shapes are also designed to provide for convenient attachment from different directions. Two common examples of structural shapes are C-channel (Figure **A**), sometimes called a U-channel, and L-shaped angle (Figure **B**) sometimes just called angle.

Protection From the Environment

The environmental forces that your materials are exposed to must also be taken into account to ensure the long term success of your projects. If your structure is made of carbon steel, you must paint the steel appropriately to resist the environment. (Remember that drilled holes should also be painted or corrosion-proofed with primer prior to assembly, so there is no exposed metal). Non-pressure-treated wood needs to be sealed to protect it from the elements. Some plastics become very brittle when exposed to acid rain or UV light.

Tool Availability

In general, select materials based on the tools you have available to you, or for those that you're willing to acquire or gain access to. However, don't feel you have to be restricted in material choices just because you don't have the tools or experience to work with them. Do a web search for makerspaces, hackerspaces, fab labs, tool lending libraries, and community colleges in your area, where you may be able to get access to the tools, as people who will train you to use them. Consider your new project a suitable excuse to learn how to handle and work with new materials, as well as use new tools.

Material Availability

Always take into consideration how much material you have available for a project. If you need more of that material to complete your project, be sure to get it BEFORE you absolutely need it. Try to avoid situations where your project is stalled for weeks because you're waiting for more material to arrive.

Most of the raw materials you work with as a maker can be found at big-box and independent hardware stores or, if you don't mind waiting a bit, purchased online. But there

I-beams and C-channels are very common structural components. In the construction business, these are found just about everywhere in every building holding up floors and ceilings. An I-beam (shown below) is simply a piece of material that has a cross section that looks like the capital letter "I." This shape resists bending forces very well at any end of the beam. A C-channel is essentially an I-beam cut down the center of its cross section so that the cross-section looks like a capital letter "C."

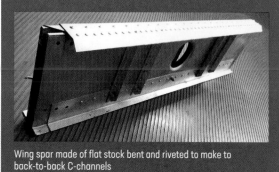

Wing spar made of flat stock bent and riveted to make to back-to-back C-channels

are other ways to source project materials. Here are a few suggestions.

REPURPOSING COMMON MATERIALS

Makers and professional engineers alike share the need to economize. When selecting materials for your projects, consider what serves your needs best in terms of the planned functions and operational limits of your design, as well as the availability and costs of the materials you could potentially use. Next time you stroll through your local big-box or hardware store, examine the materials on display – boards, metals, PVC pipe sections – for their potential use in current and future projects. Before you give in to the allure of purchasing new goodies, though, think about the materials you may already have on hand. You can save a lot of money if you are innovative and repurpose things you already have lying around your house or workshop. Besides saving money, you'll also:

- **Save time:** Save yourself a trip to the hardware store and get your parts immediately! Materials that are already on-hand take no time at all to procure. Even if the parts are not exactly what you need, you may be able to adapt something you already have to meet your needs with a minimum of processing.

- **Create less waste:** Get on the bandwagon: Repurposing is COOL! Reducing landfill waste is always a good idea,

so even if an item is recyclable or you don't have an immediate need for it (like an unused length of 2"×4" board), hang onto it for future use.

- **Free up space:** We try to make as much use of material we already have on hand as we can, simply to make room in the garage!

We always try to salvage parts and raw materials from previous projects. For example, the Adult-Size Tank's original frame, which was made of wood, was turned into an ATV/Monster Kart frame when the tank was converted to an all-metal design (Figure **C**).

Makers typically repurpose materials that they just happen to have lying around. Here are some examples of things we've found especially useful:

- **Door hinges:** We use these for pivoting parts on projects. If you want something that folds, nothing is easier to make use of than an ordinary door hinge or cabinet hinge.

- **Computer components:** Discarded computer cases are frequently made of steel, and the side panels are sheets of very useful material. We typically cut off the trim and protruding parts and use the remaining plate material. Server rack rails, which are essentially pieces

Tracking Further: Scrounging

Many makerspaces host "take apart" events. Disassembling old components and machines is a great way to learn how things work, but it's also a great way to harvest parts for future projects. While much of what we harvest is electronic components (LCDs, lasers from DVD drives or motors), we also harvest hardened steel rods (inkjet and laser printers provide excellent materials), bearings, gears that are guaranteed to properly mesh (and fit those rods), switches, glass plate, and even ABS plastic, some of which, if you are so inclined, you can shred, melt, and then extrude to make filament for your 3D printer. Don't hoard useless junk, and be selective with what you keep, but take advantage of the valuable materials in old components.

There are other forms of scrounging, too! Anything made of ABS that can be chopped up into small pieces can be fed to something like a "Filastruder" or similar filament extruder so that you can make your own filament for your 3D printer. Don't attempt to cast ABS, though, as it is better suited to being injection molded, owing to its propensity for burning.

Finally, a more extreme form of scrounging can be illustrated by our friend's practice of creating aluminum ingots from cans (shown below). He collects all the drink cans his family and friends consume, and melts them down into ingots that he casts in a muffin tray. These metal ingots can be melted again and cast into a variety of useful shapes. According to him, it takes 150 cans to make a single muffin sized ingot, and each can is comprised of approximately 50% paint by mass! Now THAT is extreme scrounging!

A stack of aluminum ingots from drink cans, cast in muffin molds

A Server rack rails turned into C-channels for later use as structural members

of steel angle, can be used for structural parts of larger structures. In Figure **A**, two pieces of rail have been joined together to make a C-channel. Some computer cases are themselves solid enough to be bolted together as-is to make a solid frame, with the hollow area as a place to run wiring, or as storage space for batteries and other bits and pieces.

- **Wood from remodeling projects:** Remodeling projects usually result in leftover lumber, most of which is destined for the trash. Salvage this leftover wood whenever possible. For example, salvaged 2"×4"×8' lengths of pine stud material became the Adult-Size Tank's drive system mounts, after pulling a few nails and cutting off cracked ends.

Look around you! What are some of the leftovers and common parts in your garage or basement that might become the basis for future projects?

Materials Breakdown for Brian's Tot-Size Tank

As a way to summarize and break down the process of material selection, let's walk through what makes up the Tot-Size Tank that we first described in Chapter 1. We will look at each module (Figure **B**) from the standpoint of which materials are used, and why Brian chose those materials. As you will soon see, most of the materials used in the tank are really quite common, low-cost, and easily-accessible (with a few exceptions).

The Tot-Size Tank is an ultimate example of "maker-style" materials selection. The materials used range from recycled parts, to purchased raw materials, to parts and components scrounged and salvaged. Recall that the tank is divided into three main sub-assemblies or modules. The modules are as follows:
- **Module 1:** Tracks
- **Module 2:** Main frame/Drive assembly
- **Module 3:** Body

In conceptualizing, designing, and building the Tot-Size Tank, Brian started with the tracks. In fact, acquiring the

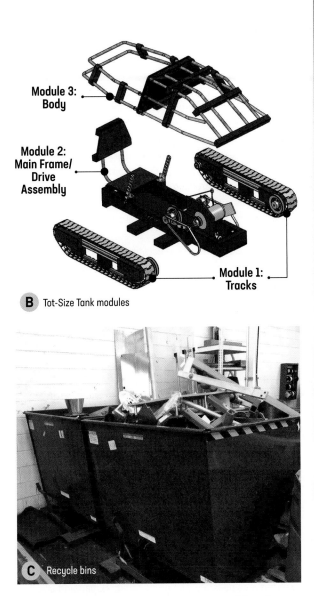

Module 3:
Body

Module 2:
Main Frame/
Drive
Assembly

Module 1:
Tracks

B Tot-Size Tank modules

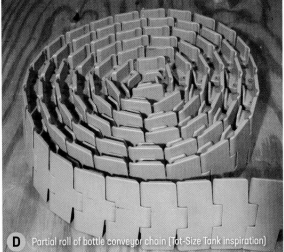

C Recycle bins

D Partial roll of bottle conveyor chain (Tot-Size Tank inspiration)

actual track material gave him the idea of building the tank in the first place.

Brian works for a company that builds and sells new packaging equipment, as well as rebuilding old, worn out, or outdated packaging machines. Once rebuilt, the equipment is like new, or in some cases better than new. When rebuilding a machine, much of/a good deal of the electrical and mechanical components are replaced, and this old stuff that they take off is thrown away, or placed in large metal recycling containers (Figure **C**).

These recycling containers are a veritable treasure trove

for materials, and Brian has been known to literally dumpster dive into these bins once in a while. Although, with regards to to the tank, he didn't have to dive for the track materials: A co-worker, knowing Brian's enthusiasm for tinkering and making, told him, "I've got something interesting for you. I have no idea what you could use it for, but I'm sure that you will come up with something."

Following him from the engineering office area to the rebuild area in the back of the shop, Brian found approximately thirty feet of 3¼"-wide bottle conveyor chain from a rebuilt machine, rolled up and destined for the recycle bin. A bottle conveyor chain (Figure **D**) is made

Staying on Track: Material Acquisition

There are a couple of important maker lessons to be learned here with respect to the acquisition of materials and components that might be useful for your projects. The first is to let it be known that you are a maker. Talk with others. Tell other people what you're working on or thinking about making. Once people know this about you, they tend to automatically think of you when they see components or materials that could be used for a project.

The second is to not be afraid to dumpster dive. (Just be sure to get permission from the dumpster's owner before you start.) Keep an eye out, and take advantage of opportunities to scrounge for materials when they arise. Often, the materials you find or are given will inspire you to think of new projects, instead of the other way around!

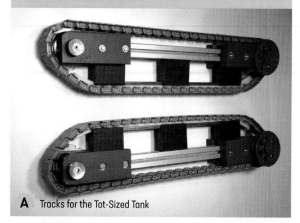

A Tracks for the Tot-Sized Tank

B Track tensioning

up of rigid plastic links, and is typically used in bottling or packaging plants to convey bottles, such as those used for soda or beer.

Immediately, Brian's thoughts went into full-scale maker mode, thinking about things that could be made from the chain. Very quickly, it became apparent that the chain would be an awesome solution for a material to use as tracks for a small tank. This is where the concept of the Tot-Size Tank began: Brian decided not to build a train, but to use as many of the train components that he had already gathered as possible to build a tank.

Though, obviously a critical component of Module 1 (Figure **A**), the chain is certainly not the only item that makes up the track module. The chain must have a structure to support it and a means to drive it.

Central to the support structure of the tracks is the "backbone." Knowing that this backbone had to be very strong, relatively lightweight and easy to attach things to, Brian quickly honed in on aluminum T-slot extrusion as the best material. Aluminum T-slot extrusion is extremely strong for its weight and the T-slots make it very easy to attach items to it, especially things that will need to be moved along the longitudinal axis of the extrusion. For the tracks' backbone, this was an especially helpful aspect of the material. The tensioning of the tracks on the tank was achieved by simply sliding the idler end bearing blocks along the backbone via the T-slots (Figure **B**).

The next challenge was to determine how to support the chain and, by extension, the entire weight of the tank and rider. The tank design was always intended to be as simple as possible, so complicated suspension and wheels to support the track chain were not the way to go. The support material needed to be stiff enough to support the chain over the entire contact area between it and the ground, while also permitting the chain to slide along it with a minimum amount of friction. Again, from the world of conveyors, a conveyor guide rail (Figure **C**) was the ideal material to support the chain and provide for a bearing surface.

C Chain guide rail

Staying on Track: Bending the Ends of the Guide Rail

Sometimes you come across materials that, with a little modification, can be made to fit the requirements of a project. For the Tot-Size Tank, the ends of the rails needed to be bent upward to offer smooth, rounded corners with angled tips, providing a lead-in and lead-out for the chain.

The guide rail was cut to the overall length, the metal trimmed on each end, and the UHMWP (plastic) bent upward on each end to provide the proper approach angle of the chain to the idler and drive sprockets.

To trim the metal from the guide rail, the plastic was slid out from one side of the metal channel, leaving approximately 6" of stainless steel channel exposed on the other end. The exposed stainless steel channel was then cut off, and the remaining stainless steel channel was slid back and centered on the plastic, leaving approximately 3" of bare plastic on both ends.

To bend the plastic ends, one end of bare plastic was clamped in a vise. The plastic close to the stainless steel channel was carefully heated using a propane torch (taking care not to melt the plastic, just soften it!) and bent toward the stainless steel channel to approximately 20°, as shown below. The same process was used for both ends. *As a word of caution, always heat plastic in a well-ventilated area. Most plastics will give off noxious gases when they are heated, melted, or burned.*

TIP: Slightly over-bend the plastic and hold it in place until it is cool or, pour water onto the plastic to cool it even faster. Once cooled and released, the plastic will spring back to the proper angle.

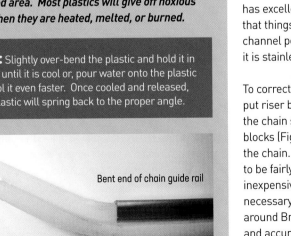

Bent end of chain guide rail

D Riser blocks

E Bearing block

The rail is made up of a length of UHMW (Ultra High Molecular Weight) polyethylene plastic that is keyed into a channel made out of stainless steel. This type of plastic has excellent wear properties and good lubricity, meaning that things slide easily along it. The stainless-steel channel portion of the rail makes it very stiff and, because it is stainless steel, corrosion resistant.

To correctly maintain the track chain's geometry, Brian put riser blocks (Figure **D**) between the backbone and the chain support guide rail. Also, he added bearing blocks (Figure **E**) to support the drive and idler ends of the chain. Both applications required the block material to be fairly strong, easy to cut and drill, and relatively inexpensive. Standard 2"×4" lumber had all of these necessary properties, and there was plenty of it lying around Brian's workshop. He used a chop saw to quickly and accurately cut the blocks, both a common spade

B Polycarbonate sprocket (no hub)

C Machined hub

and Forstner bit (see more on Forstner bits in Chapter 4) to drill the through-holes and counter-bores (to accept the axles and bearings), and a router to cut a channel to accommodate the chain guide rail in the bottom of each of the riser blocks.

The bearings that support the drive and idler shafts are simple, ordinary single-row ball bearings, but they were not obtained or contained in a normal, "buy-them-from-the-store" manner. At Brian's workplace, rollers are one of the items that are always replaced in rebuilt machines. So there is a ready, steady supply of used rollers in the recycle bin. The rollers consist of a hollow metal tube with ball bearings pressed into both ends, and a slightly longer steel shaft going through the center. Figure **A** shows several of these roller assemblies mounted in a machine. Rather than purchasing new bearings, Brian simply removed the roller's shaft and cut the tubing near the bearing ends from a few of these rollers, generating short sections of the hollow tube with a bearing contained within. The outside diameter of the roller tube matched a standard sized Forstner bit, so drilling the counterbored holes in the bearing blocks was simple: The recycled bearings pressed nicely into the counterbored holes. The roller shafts were obviously already a proper size to fit the bearings, so they were also recycled as the track idler and drive shafts. The main live, drive shaft was directly linked to the motor.

To drive the tracks, Brian needed a sprocket that fit the particular bottle conveyor chain he was using. Knowing the make and model of the bottle conveyor chain, it was quite easy to find sprockets that would fit the chain, but they were not the exact size that Brian needed for the tank, and purchasing the sprockets would have been expensive. The tank needed a total of four sprockets (two per track: one idler and one drive), and that would cost nearly $300, assuming that the design could be modified to use one of the sizes available! Fortunately, the chain vendor had solid models (also known as 3D CAD files) of the available sprockets online. So after downloading one of the models, Brian was able to utilize the vendor's geometry to design a sprocket with the proper scale, pitch diameter, and number of teeth (these terms are discussed further in Chapter 8) that could be made using simple, common workshop tools. The material for the sprocket had to be very tough, but not abrasive enough to damage the chain, and it also had to be easy to cut and drill. He

found some old, broken plastic machine guards made out of ⅜" thick polycarbonate, also known by the trade name Lexan. In a thicker size, polycarbonate can be made into bullet-proof windows, so this material is plenty tough enough to serve as sprocket material (Figure **B**). Additionally, polycarbonate is quite easy to cut and drill: Unlike its plastic cousin, Plexiglass, it is not brittle and will not crack.

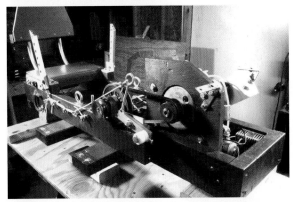

D Sprocket and hub

The tank required hubs that would support the sprockets and provide a means by which to lock them to the drive and idler shafts. As such, they needed to be relatively light and strong, and able to be tapped to accept set screws to lock against the shafts. These hubs also needed to be fairly thick, to distribute the load over a reasonable area of the shafts. After doing some research, Brian decided to purchase round 6061-T6 aluminum bar cut to the desired thickness of the hubs. Using his small, inexpensive metal lathe, Brian machined the hubs (Figure **C**) from the purchased aluminum bar to fit the hand-made sprockets. Figure **D** shows the Lexan sprocket bolted to the aluminum hub.

When designing the main frame of the tank (Figure **E**), Brian had to consider several important specifications or design parameters. The main frame had to provide the structure on which to attach the tracks. It also had to make room for and support the various drive pulleys, drive motor, and control handles. This frame also had to accommodate and provide a way to protect Brian's son from the mechanics and electronics of the drive system as he operated the tank (Figure **F**). In other words, this frame had to provide a way to mount the body.

As he did with Module 1, Brian selected the materials for Module 2 to balance function, cost, ease of manufacturing, and aesthetics. The material of choice for the main frame structure was standard 2"×4" lumber, which is a strong material that is easy to cut and drill, making it easy to mount components anywhere to the frame. In retrospect, it might have been better to build the frame out of aluminum T-slot extrusion, as he did for the backbone of the tracks. The T-slot extrusion has a much larger stiffness-to-weight ratio than standard 2"×4" lumber, so it provides the same rigidity as a 2"×4" with a much smaller cross-sectional area and weight. The downsides of using T-slot extrusion are that it is quite expensive, and you are restricted by the location of the T-slot as to where you

E Main frame and drive

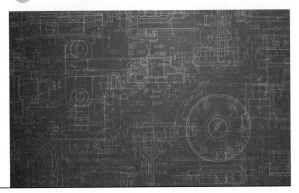

F Test fit with Evan

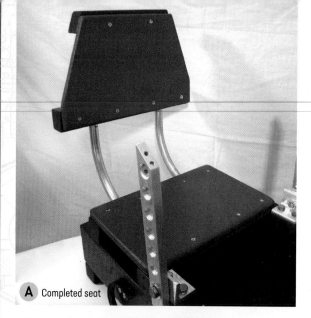

A Completed seat

B Motor mount

C Tensioner and motor support

D Brake belt

can mount components to it. However, if you can spend more money on frame materials and you plan out your components and the mounting thereof, aluminum T-slot extrusion is really the way to go.

In order to stiffen the overall frame structure, Brian used ⅜" thick plywood for the floor of the tank. Plywood is an excellent material for skinning structures and for increasing the overall stiffness of frames. It is very strong, not too expensive, and easy to cut, drill, and shape.

The plywood also stiffened the seat structure, and it served as the bottom and back surfaces of the driver's seat because it was nice and flat and easy to work with. The initial surface of the plywood was quite smooth, but with a little sanding became very smooth, with a nice, rounded edge, as seen in Figure Ⓐ. This was an important material consideration for the seat, as a rough seat edge might cut the driver.

The main motor mount was also constructed of plywood: in this case, ½" thick for added stiffness and rigidity. Brian went to a hardware store, where he purchased a bracket made from zinc plated carbon steel. He attached this to the motor mount, as seen in Figure Ⓑ, where it became one of the two front mounting points for the body (Module 3).

In early stages of the tank, the motor was supported by plywood at only one end, a method termed a **cantilever support**. Later, when Brian became concerned that the motor mount would not provide a rigid enough support for the motor by itself, he added a length of ¼" thick aluminum flat stock to support the other side of the motor, as seen in Figure Ⓒ. This support also provided a mounting point for the other front body mount bracket.

Each track had a pulley attached to its idler sprocket shaft that worked as a brake. He fixed a length of v-belt to the frame at one end and tied it to the control handle at the other end, so that when the driver pulled back on the control handle, that pulled the V-belt (Figure Ⓓ) against the track brake pulley (Figure Ⓔ), thereby generating friction, and thus generating a braking force. This is a good example of material being repurposed and used in a slightly different way than originally intended.

The control handles (Figure Ⓕ) and drive belt tensioner pivot arms (Figure Ⓖ) were all made out of ½" thick

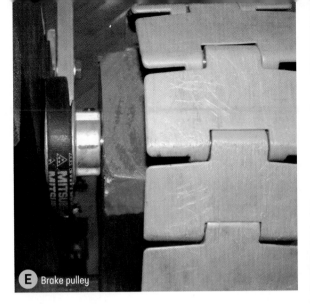

E Brake pulley

Control handles

F

G Drive belt tensioner

scrap 6061 T-6 aluminum bar stock that was lying around Brian's workshop.

6061 T-6 aluminum is very strong for its weight, easy to cut, drill, and tap. Also, it looks it looks quite nice. Aluminum will dull due to surface oxidation, but it will not rust like carbon steel. You can also buy aluminum that has been anodized: run through an electrochemical process that deposits an inert oxide layer of material on the surface of the aluminum. In most cases, aluminum extrusion will be clear-anodized, but it can sometimes be obtained in other colors, with black-anodized extrusion being very common. The downside of using aluminum is that it can be quite expensive. Plywood or a hardwood dowel, such as oak, can often easily be substituted.

The drive handles and drive belt tensioner pivot arms were attached to the frame via a bracket that allowed them to pivot relative to the frame. Brian used 1½" × ¼" thick aluminum angle stock. The cross-section of this material provided for a very strong structure in a fairly compact form. Carbon-steel angle would have worked equally well for this application. It would have been less costly but would have required painting to prevent rust.

Earlier, we mentioned that the seat was stiffened with plywood at the bottom and back. But the underlying structure of the seat was made of two interesting, yet very common materials.

The two shiny tubes connecting the lower portion of the seat to the seat back (Figure **H**) are are simply lengths of ½" electrical conduit (also known as EMT). EMT is

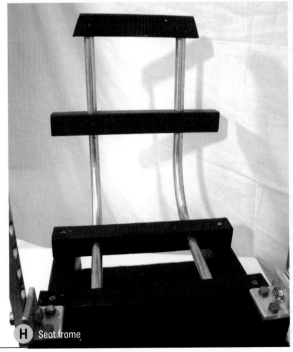

H Seat frame

Tot-Size Tank body

B Body showing hardware cloth under structure

C Tot-Size Tank dash

a thin-walled steel tubing that has been galvanized for corrosion resistance. The wall thickness for ½" EMT is approximately ³⁄₆₄". This is a great material for making light-but-rigid structures. It is simple to cut, can be bent easily with a hand tube bender, is corrosion-resistant, and is very inexpensive. Brian used standard 2"×2" lumber to tie the EMT tube structure together. It was the perfect size to accept the EMT tubing. He drilled holes through the 2"×2"s to allow the EMT tubing to pass through the wood, and the tubing was secured to the 2"×2" cross-members with woodscrews.

The construction methods and materials for the body (Module 3), shown in Figure A, were largely based on the takeaways from building the seat structure.. The main structure was made from ½" EMT tied together with standard 2"×2" lumber, making a very rigid yet relatively light body. The front portion of the body had to protect both the driver and anyone around the tank from the drive pulleys and belts, though it was important to Brian that the drive and electrical components still be seen: They looked kind of cool and, from a troubleshooting standpoint, it would be good to be able to safely view this area during operation. Screen material made of galvanized steel, called "hardware cloth," was installed under the main structure to meet these requirements

(Figure **B**). Hardware cloth is not too expensive, and it is easy to cut and shape (use gloves!).

The only other material used in the construction of the body was ¼"-thick plywood, used for the dash (Figure **C**). It was easy to transfer a pattern for the cutouts needed for the instrument displays, lights, and buttons, and simple to cut them out using various size drill bits and a hand saw.

Materials Breakdown for Samer's Adult-Size Tank

When Samer set out to build his Adult-Size Tank, he set a series of rules (or specifications) around his build, one of which was to use as much material as he happened to have, or could source himself. Samer's tank was going to need to support an adult weighing as much as 250 lbs, as well as 80 lbs of batteries, a pair of 20 lb motors, and, of course, the frame and assorted bits and pieces. He therefore chose to build his tank out of his GoKart frame (which was made of 2"×4" wood, with a broken cafeteria chair for the seat) and to machine the wheels out of plate aluminum.

Samer borrowed a drawing from Brian, and then sent off his modifications to someone he looked up online to have the wheel cut via a water-jet cutter. The wheels were made of cut plate aluminum; nuts, bolts, and washers bought at the hardware store; a bronze flanged fitting epoxied onto the wheel as a bearing; and another large bolt used as the axle (Figure **D**).

Samer used the same design, with a smaller-sized central hole as his drive sprocket, and "appropriated" a hub from a robot project to mate it with the motors (Figure **E**).

As Samer continued to develop the tank, he came up with a variety of ways to secure the road wheels (i.e., the wheels that the tank tracks roll on). In the end, he bolted the wheels directly to a set of pillow block bearings. Figure **F** is a picture of a pillow block mounted as part of a transmission system.

While the Adult-Size Tank was a success — it moved a reasonably-sized adult at a reasonable pace — it did not do well against large obstacles. Tracked vehicles, when encountering an obstacle head-on, will climb over the object as long as it is not taller than the leading (or trailing) section of the track. There is also a limit to what

Adult-Size Tank road wheels on a test sprung bogie (wheel)

D

E   Adult-Size Tank drive sprocket and hub

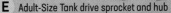

F   Transmission components using pillow block bearings

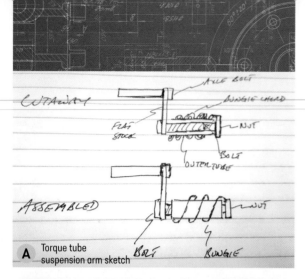

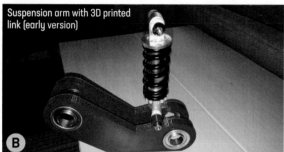

A Torque tube suspension arm sketch

B Suspension arm with 3D printed link (early version)

C Suspension arm with 3D printed link (late version)

D Sprung bogie concept

they can clamber over. The obstacle can also destabilize the tank, making the tank more likely to roll over. To get around this, the tank needs suspension, as Samer added in the Adult-Size Tank version 2.0.

In the next figures, you can see the evolution of his suspension arm, from a torque tube design (Figure **A**) to a trailing link design (Figures **B** and **C**) to a sprung bogie (Figure **D**). Notice that again, he used multiple materials: plastics, metals, and components derived from all over. The threaded rods used to hold the printed parts together came from a DIY store, as did the nuts. The white and blue plastic parts were 3D printed. The bearings were purchased online. (Before Samer designed the arm, he looked for bearings that would allow him to re-use his existing road wheel bolts and pillow block bearings, which subsequently drove the design.) The shock absorber came from a child's BMX bicycle, one of several he found for $10 online per pair. The wheels were skateboard wheels, as were the bearings for them.

The point of all this is that when you consider an "assembly," try to encapsulate each sub-system in your design. See if any parts can be derived from other commonly available things, and don't be bashful about revisiting your concept multiple times. As hard as we try to design on paper, there is always going to be some degree of experimentation. This is what we call "prototyping."

Conclusion

Through this chapter, we hope that you have found ideas for materials to use in your own projects, and also learned learned to think like a maker when selecting the materials for your projects. You may not run across something exotic like conveyor chain, but the odds are good that there are things around you that you can repurpose into interesting new projects. How can you repurpose things to meet certain design requirements? Think outside the box when it comes to materials for your project.

Tracking Further: Sources of Materials

In this book we mention several places where you can find materials, but one of the easiest places for this is right in your own home. Some useful materials I have found in my own home include:

- **Plastic milk jugs:** Most are made from polyethylene, which can be heated and formed into a variety of shapes, cut with scissors, and glued with CA adhesive. Some potential uses: small patches for holes, makeshift funnels, insulating washers, diffuser lenses for lights.

- **Plastic wrap and plastic sandwich bags:** Stretchable, make good air and liquid barriers (though not against significant pressure). Can be used for diaphragms for horns (as presented in the air horn project in Chapter 2), masking material for spray painting, and many other of other uses.

- **Wax paper:** Does not stick to most anything. Can be used to prevent unintended adhesion when dealing with a messy glue, like polyurethane, when laminating things together.

- **Aluminum foil:** Easily shaped to fit tightly around objects. Useful as a masking material for paint. Cosplay makers appreciate it as a way to mimic genuine metal items for use on costumes, without having to actually make them from metal. They use foam or plastic to create the item's shape, and then cover it with aluminum foil to make the item appear metallic.

- **Paper clips:** Specifically, the kind made from steel wire. These can be bent and used in a multitude of ways: clips, small linkages, tie wire, hooks, etc.

- **Rubber bands:** great for temporarily clamping smaller items for gluing, or for a quick test assembly. They serve well as tension springs for smaller projects, and they can easily be grouped together to adjust the spring force.

- **Toothpicks:** Specifically the round kind. They're excellent as glue and paint applicators in tight spaces. They can also be used as axles or wooden dowels in very small projects.

- **Old toys:** Nearly every household has some old toys lying around. The possibilities are almost limitless here. You can harvest all manner of screws and other hardware, bearings, wheels, and even electronics, lights, and lights and speakers from some of the more complex ones. I'm always grabbing battery boxes and battery chargers out of old toys to use for other things.

- **Old appliances:** Just like toys, but usually stronger. You can get bearings, shafts, gears, motors, and lots of other mechanical parts from these.

- **Old electronics:** Again, a treasure trove. Old audio equipment usually has lots of nice pushbuttons, potentiometers, numeric displays, control knobs, and other goodies. Old computers, especially, have power supplies and cables that come in handy. Don't forget the stylish metal cases they comes in for use as enclosures, or metal panels on a future project. Make a habit of keeping those unneeded plug-in power adapters (wall warts) that nearly everything uses now. Brian has a cardboard box full of these for future use.

- **Old clothes:** Not only for rags in the workshop, but snaps, buttons, zippers, belts, and even old pockets can be repurposed.

- **Old containers:** These could be many things: that old Tupperware container under the sink with the missing lid, an old mint tin, anything made of plastic or metal that has fallen into disuse. The plastic ones are great for making recycled power supply touch-safes. And let's not forget they're great for storing all of the other materials on this list, once you start accumulating them.

Even though some of these are already mentioned, there are quite a few other potential sources for parts and materials, if you know where to look:

- **Home improvement and DIY stores:** These are probably your primary go-to places for materials. You'll find a wealth of structural materials (wood, metal, plastic, hardware cloth, etc.), fasteners of all kinds, PVC pipe and pipe fittings, conduit, miscellaneous hardware (bearings, pulleys, cable and hooks), electrical components (switches, lights and wire), and of course tools.

- **Auto parts & farm supply stores:** These are good places for all kinds of things,as structural materials (mostly metal and plastic), fasteners, rubber hose, wheels, tires, casters, belts, miscellaneous hardware (hose clamps and pins), electrical components (switches, lights and wire), and some specialty tools.

- **Hobby shops:** These are great sources for styrene plastic, in sheets and other shapes, as well as small sizes of brass and aluminum tubing, balsa wood, paper tubes used in model rockets, and fine wire used for making jewelry.

- **Craft supply stores:** Here you can find many of the same things as in hobby shops, plus fabric, ribbon, cardstock, foam core boards (laminated boards made from foam with paper on both sides), picture matting (very heavy paper cardstock), and one of my favorites; Coroplast (corrugated plastic sign material). Coroplast is great for building very lightweight enclosures inexpensively.

- **Pull-apart auto wrecking yards:** This is where you can find the heavy-duty stuff for bigger projects. Pulleys and bearings are obvious choices. But you can also find all manner of metal brackets and parts that can be repurposed. Windshield wiper motors, power window motors, and door lock solenoids are abundant. And don't forget hose clamps, rubber hoses, wire with good weatherproof connectors, plastic wire loom, switches, knobs, relays, lights, horns, speakers, rubber O-rings and seals, and any size and shape of fastener you can imagine (and a few you probably haven't).

- **Yard and garage sales:** Everything on the "at home" list above is also available at other homes (when the owner is selling; don't steal). If you're lucky, you might score some second-hand building materials quite cheaply, or even free. I'd include flea markets and other reselling venues in this category.

- **Dumpster diving:** This can be tricky. But if you ask permission from the owner, sometimes you can get useful materials that are considered waste by local businesses. For example, a local cabinet shop may throw out wood scraps that you may find useful. But beware: Many businesses will say no just to avoid any possible liability. If you do get permission, be courteous, be safety-conscious moving around stacks of material, and avoid items used with hazardous or flammable chemicals. And please don't leave behind a mess, or you (and other makers) may not be allowed another visit.

- **Online:** It wouldn't be the 21st century without a reference to internet sources for materials. The list of buy/sell/trade sites is growing all the time. At the time of this writing, I like Craigslist, Facebook Marketplace, Offerup, Letgo, and eBay. But options vary with geography. Beware of shipping costs if you buy something heavy that's far away!

This is not an exhaustive list by any means. But it should get you started.

One last point I want to emphasize is that while making can be a solitary activity, you shouldn't be afraid to network with other makers. Aside from the wealth of experience, ideas and creativity you can tap into, others may have tools, skills and materials that you can barter for, or obtain reasonably inexpensively. —*John Manly*

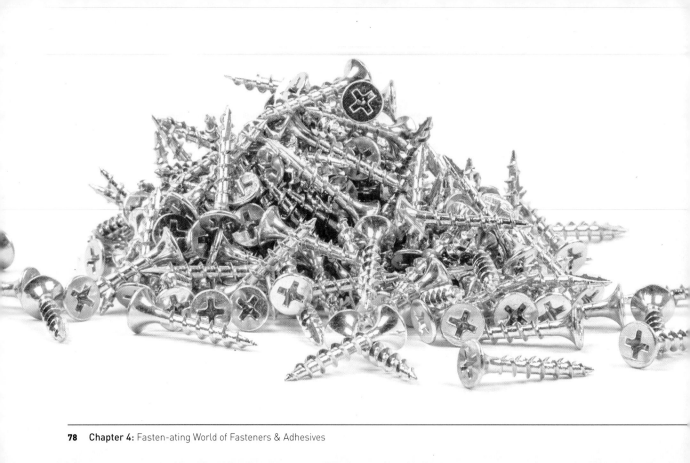

4 Fasten-ating World of Fasteners & Adhesives

In the manufacturing phase of your project build, you will need to have a means by which to fasten parts and components together. This is where fasteners and adhesives come into play. For the purpose of this discussion, we classify fasteners as physical devices used to attach or fasten materials, parts, and components together in a non-permanent way. Adhesives are generally defined as a liquid or thin film that bonds two materials or parts in a more permanent fashion.

Types of Fasteners

Fasteners come in all sorts of sizes, types, and materials offering functionality that goes well beyond the proverbial nut-and-bolt way of keeping things together. Consider the simple screw: It's nothing more than a spiral wrapped around a shaft that is either tapered or flat at one end, and a head at the other that can tightly hold two or more parts together. Fasteners are available in both metric and US Customary System units of measure, which can sometimes cause confusion. We could write several chapters on all the various fasteners you see just around you on a daily basis. In this chapter we cover several fasteners that we, as makers, have found to be most useful for a wide range of projects.

Hep Svadja

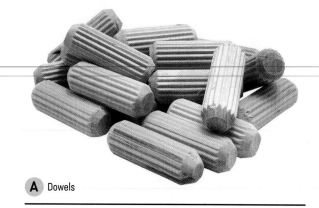

A Dowels

We are sure you will find the array of fastener types quite dizzying. Fasteners have been around for millennia and run the gamut from the ridiculously simple to seriously complex.

DOWELS (Figure **A**): These are pegs without a head that makers can pound into a hole. Wooden dowels, for example, when coated in wood glue, expand into the hole as the glue cures, and secure parts together very tightly. Dowels are typically made of wood, and sometimes have longitudinal ridges that dig into and "grip" the sides of the hole, trapping glue in between them.

B Nails

NAILS (Figure **B**): Nails have a tapered end that forces apart the material they are being driven into. The material then contracts around the nail, holding it in place. The material exerts a greater force on the nail as the nail is driven deeper.

RIVETS (Figure **C**): You will mostly encounter "blind" rivets. Also known as "pop" or "pull" rivets, these rivets have a shank that breaks as you pull them with a special riveting tool, thus securing the parts by compressing and deforming the end of the shank. Like screws, these come in pan head (a round, domed head) or countersunk (V-shaped head). You might also encounter plastic rivets that can be put in and taken out by hand but are really "low" grip fasteners.

C Pop rivet

PINS (Figure **D**): These are similar to dowels, but are typically made of metal. A common example of a pin is a **roll pin**. A roll pin is typically made from flat spring steel rolled into a cylinder. When a roll pin is driven into a hole, the pin collapses slightly, thereby jamming itself into the hole.

SCREWS: Screws can generally be divided into three types: wood screws, sheet metal screws, and machine screws. Screws with tapered threads are designed to screw into a softer material, while screws with straight threads are designed to be used with a nut, or thread into a matching threaded hole. Additionally, some wood or sheet metal screws are designed with a small drill-bit type point at the end, and referred to as "self-tapping" or "self-drilling" screws.

D Pins

- **Wood Screws** (Figure **E**) have tapered threads, and are designed to be threaded into a plain hole in wood, so that the wood deforms around the threads of the screw to hold it in place. They generally have a non-threaded portion near the screw head that is intended to fit loosely in a slightly oversized hole in one layer of material, with the threaded part in a second layer, holding the two layers together. Wood screws also frequently have a flat head intended to fit into a countersunk hole in wood (or designed to make one as the screw is driven into the wood). Larger sizes of wood screws, used for heavy structural connections, are called lag screws, or even lag bolts.

- **Sheet Metal Screws** (Figure **F**) also have tapered threads, and are designed to be threaded into softer material, allowing the material to deform and hold the screw tightly in place. Sheet metal screws look similar to wood screws, but are made to hold much thinner layers of material together. They tend to have smaller threads, and they usually have a round or hexagonal head that stands proud of the material they're driven into. Also like wood screws, machine screws are not used with nuts or pre-threaded holes.

- **Machine Screws** are designed to be threaded into a nut or threaded hole in another part. Their threads are not tapered. Larger sizes of machine screws are called **bolts** (Figure **G**).

- **Carriage Bolts** (Figure **H**) have a domed head with a square portion below the head. This type of bolt is designed to be used in applications where a smooth, low head is required for aesthetic or safety reasons. An example of when this type of bolt is used is when building children's playground equipment. If children brush up against the smooth head of a carriage bolt, they will not hurt themselves. When a carriage bolt is used to secure wooden parts, the square portion of the bolt is designed to be pulled or driven into a clearance hole drilled for the bolt. The square feature "bites" into the material around the clearance hole, thereby keeping the bolt from rotating while the nut is secured to the other end. This type of bolt can also be used to secure metal parts together. For this type of application, the top layer of metal to be secured generally will have a square hole to receive and engage the square part of the bolt.

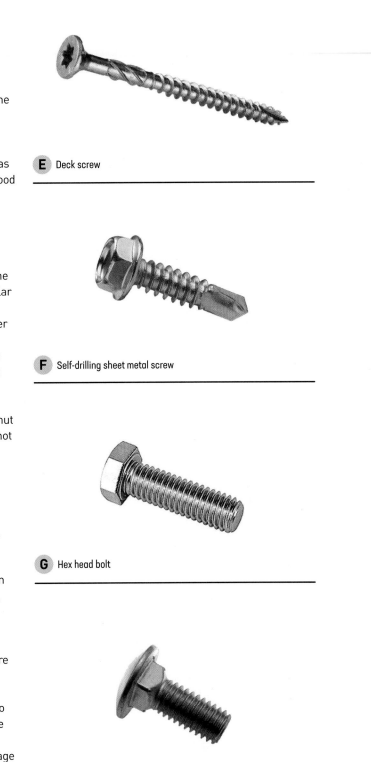

E Deck screw

F Self-drilling sheet metal screw

G Hex head bolt

H Carriage bolt

A Shoulder bolt (also called a stripper bolt)

B Eye bolt (left), eye lag (right)

C Hanger bolt

D Hex nut

E Nyloc nut

- **Shoulder Bolts,** also known as **stripper bolts** (Figure **A**) have a very precise shank or smooth portion of the bolt, intended to be used as a pivot. The end of the bolt is threaded with a thread diameter less than that of the shank. This smaller-sized thread creates a step at the bottom of the bolt that the bolt seats against when it is tightened down to the face of a material.

- **Eye bolts and Eye Lags** (Figure **B**) have a circular ring on one end and threads on the other. Eye bolts have a machine screw thread while eye lags have a wood screw thread. This type of fastener is used for securing rope or chain.

- **Hanger Bolts** (Figure **C**) have a machine screw thread on one end and a lag or wood screw thread on the other. This type of fastener is used when a machine-screw-threaded stud is required to protrude from a wooden part.

NUTS: Like screws, there are plenty of variations:

- **Regular Nuts** are typically made of metal, with a threaded hole running through the middle. The outside of the nut has a specific geometry by which the nut is intended to be gripped and rotated. It is common for the outside geometry to be a hexagon, called a "hex nut" (Figure **D**), but in making you will occasionally come across nuts that have a square outside geometry.

- **Lock Nuts**, sometimes called nyloc nuts (Figure **E**), are simply regular nuts that have a piece of nylon (typically) embedded in the hole that grabs the threads of a screw. This prevents the nut from working itself loose over time. This type of nut is quite handy for applications where a bolt should be mechanically secured to a component, but not completely locked in place. A typical example of this is when using a bolt as a pivot for something else, such as a lever. You do not want to completely lock the lever down with the bolt used as a pivot or axle, but you also do not want the lever to become loose or unsupported. This is where a nyloc nut can be quite useful. Using a nyloc nut, the bolt being used as a pivot can be secured in place without being fully tightened, thereby allowing the lever to rotate about the bolt.

- **T-slot Nuts** (Figure **F**) are similar to regular nuts in that they are also typically made of metal and have a tapped hole through the middle. However, the outside geometry of a T-slot nut differs significantly from that of a regular nut. The stepped outside shape of this type of nut is designed to engage in a part or material specifically designed to accept it. Aluminum extrusion is one example of a material designed to accept T-slot nuts, as shown in Figure **G**.

- **Castle Nuts** (Figure **H**) have notches in them so that a pin (commonly called a **cotter pin**) can be passed through the notches and a hole in the screw shank. This prevents the nut from rotating and working itself loose.

- **Rivnuts** or rivet nuts (Figure **I**) are essentially rivets that have a threaded hole in them to accept a screw. Rivnuts are great for securing panels in sheet metal (Figure **J**).

- **Inserts** come in a variety of forms (Figures **K** and **L**) suited to specific materials. These are similar to rivnuts in that they provide a threaded hole, but can be put into plastics, wood, and composites. Inserts for wood are commonly called T-nuts (as opposed to T-slot nuts), and get hammered into a hole in the receiving material. Thermoplastic inserts go into acrylic and similar materials by applying heat to the brass insert which melts the material around it around it. Once cooled, it forms a solid mechanical bond between the insert and the plastic.

WASHERS: Washers are important parts of mechanical fastening systems. A washer is essentially a thin disc with a hole in the middle. A **plain washer** (Figure **M**) is typically used to spread the load that a bolt exerts on the material it is fastening together. This is done to prevent the bolt head from deforming or pulling through the material.

Some washers are designed to help secure the bolted joint from working loose due to vibration. This particular type of washer is known as a **lock washer**. There are many types of lock washers, but the most typical type that you will use in your maker projects is a "helical" or "spring" lock washer (Figure **N**), which is essentially a plain washer that has been cut and sprung out of plane.

F T-slot nut   **G** Aluminum T-slot extrusion with T-slot nut

H Castle nut

I Rivnut   **J** Installed rivnut

K T-nut wood insert   **L** Thermoplastic insert

M Plain washer

N Lock washer

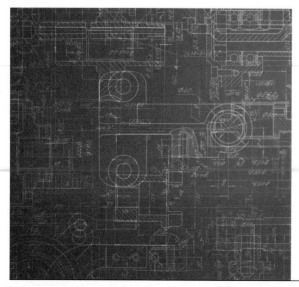

Hex Bolt
Lock Washer
Plain Washer
Plain Washer
Hex Nut

A Typical bolted joint

The way this type of lock washer works is that when a bolt exerts a load on the lock washer, it forces the washer to flatten out, causing the washer to dig into the material being fastened. It also provides some axial force on the bolt by virtue of being forced flat. In other words, the lock washer tries to spring back. The forces generated by both digging-in and springing-back help to keep the bolt from rotating loose. In most applications requiring a lock washer use plain washers as well. Figure **A** shows a typical bolted joint that has both a lock washer and plain washers. Note that only one lock washer is required for a joint such as this. However, two plain washers are used to distribute the load over a larger area of the materials being joined.

Threaded Fastener Units of Measure

All threaded fasteners are defined by their diameter, sometimes shown as "Ø", and the pitch of their threads. **Thread pitch** is a measure of the spacing between the threads. For some fasteners, such as screws, a length measurement is also specified.

A metric screw might be specified as M5×0.8, 20mm long, or M5×0.8×20. In this case, the M5 corresponds to the nominal size of the screw being 5mm. We use the term nominal because it is close to, but not exactly 5mm in diameter. The 0.8 corresponds to the thread pitch, which for metric fasteners is the distance in millimeters between each thread. The smaller the pitch for metric screws, the finer (physically smaller) the thread. Finally, the length is also specified in millimeters: 20mm in our example.

In the United States, parts such as fasteners are often sized using the US Customary System of measures or units. But while the specifications look a little different than metric screw specification, they are used the same way. A US Customary System screw might be specified as ¼"-20, 1" long, or ¼"-20×1. In this case, the ¼ corresponds to the nominal size of the screw being ¼ inch (0.25in). The 20 corresponds to the number of threads per inch (TPI). So, opposite metric threads, a larger number in pitch for US-system screws means a finer thread, and a smaller number means a larger (coarser) thread.

Tracking Further:
Screw and Bolt Hole Types

Most of the time, when assembling parts or components of a project together with fasteners, you will need to drill a hole of some sort in one or both mating parts. Figure **B** shows four common examples of hole types used with fasteners.

Section 1 in the figure shows the typical hole configuration required when using a countersink, flat head screw. This type of screw is used when you need the head of the screw to be flush or slightly below the surface. Note that the figure (Section 1) shows three different, distinct hole geometries in this one example. The top, angled area is there to provide a place for the screw's head to sit when it is tightened. This is known as a countersink. There are two main methods to achieve a countersink. You can use a countersink bit in a drill press or hand drill to bore a countersunk hole. A countersink bit is shown in Figure **C**. The other way is to simply let the screw head machine its own countersink. If you look closely at a deck screw (Figure **D**), it likely has small ridges or flues along the lower, angled portion of its head. These ridges serve to help cut away material, so that the screw forms its own countersink as it is turned and thus tightened.

The hole directly beneath this countersink is a clearance hole. This clearance hole is slightly larger in diameter than the outside diameter of the screw. Note that the clearance hole only goes through the top layer being fastened. This is so that the screw can pass completely through this layer of material and only engage the lower material being fastened. This allows the screw to apply a much larger compressive force on the upper material only. You may have had the experience of trying to hold together two pieces of wood by hand when using a wood screw to secure them to each other. Without the clearance hole in the upper piece of wood, you probably found that when the screw bit into the next layer of wood it tended to push the two pieces apart. This is a good example of of a problem that is solved with a clearance hold in the upper piece of wood.

Section 2 shows how to place a non-countersink bolt beneath the surface of the part being bolted on. A large hole with a flat bottom is drilled partway into the upper layer of material, deep enough so that the entire bolt head and any washers are even with or lower than

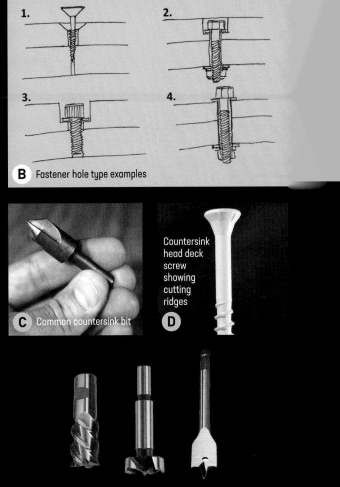

B Fastener hole type examples

C Common countersink bit

D Countersink head deck screw showing cutting ridges

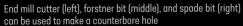

E End mill cutter (left), forstner bit (middle), and spade bit (right) can be used to make a counterbore hole

counterbore can be made using an end mill cutter, spade bit, or Forstner bit (Figure **E**). Each of these bits creates a hole with a flat bottom. (Note that spade bits and forstner bits work well with wood and plastic, but for adding a counterbore to metal, an end mill should be used.) Considering that a nut and washer are used to secure this type of bolt hole uses a clearance hole through both layers of material.

Section 3 of the figure also shows a counterbore hole application. In this case the lower material has been tapped or threaded to accept the bolt.

Section 4 shows a simple clearance or through-hole. Again, this is a hole slightly that is larger than the outside of the screw, drilled completely through both parts or layers to be fastened together. This is the simplest type of bolt hole. In most cases, this is the type of hole used in a

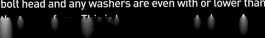

Just to complicate matters further, smaller US Customary System screws (below ¼") don't use fractional inch dimensions. They use a numbered size designation, sometimes called "gauge," starting with #12 and decreasing to #0 and even even smaller, as you can see in the screw gauge reference table (Figure Ⓐ). So, don't be surprised if you see a smaller screw specified as, for example, #6-32, ¾" long.

Earlier, we introduced the term "nominal," which basically means "what we call it" or "approximately." This is just to simplify how we talk about fasteners. A nominal M8 screw is slightly smaller than 8mm in reality, and an M8 washer has a slightly larger inside diameter or hole. This allows them to be easily assembled, while allowing for tiny variations from one part to the next. But it would be painful to refer to these fasteners by their actual sizes. So we use the nominal size to keep it simple. The exact diameter of these screws isn't as important as understanding that screws, nuts, and washers of the same nominal size will usually work together. The exception to this is when the thread pitch of two fasteners is different. For example, a #10-24 nut will not thread onto a #10-32 screw. Likewise, even though some metric and US Customary fasteners are very similar in diameter (M6 and ¼", for example), their thread pitches are different enough that they won't work together. In other words, if you have a ¼"-20 tapped hole you can only use a ¼"-20 screw. Even though the M6 screw's diameter is 0.236", which is very close to ¼", you cannot screw it into and ¼"-20 tapped hole. In this case, the diameter is not the limiting factor; the pitch is the reason they don't work together. It's important to realize that these parts need both the same size and pitch to work together. That said, these two systems of fasteners are not interchangeable if any of the holes are tapped. Though the diameters may be quite close, they are not the same pitch and are therefore not interchangeable.

In the case of clearance (i.e., untapped) holes, metric and US Customary sizes with very close diameters can be used when fastening components together. For example, if you have a clearance hole intended for a ⁵⁄₁₆" diameter screw, you can use either a ⁵⁄₁₆" (0.313") screw or an 8mm (0.315") screw since they are very close in diameter. Likewise, a screw intended for a ¼" clearance hole works with both a ¼" (0.250") screw and a 6mm (0.236") screw.

Be aware that there are two different, general classifications for thread pitch. We use the term "coarse threads" to refer to larger threads used for common screws, and the term "fine threads" to refer to smaller threads used for special situations, such as higher accuracy parts. So, in the above mismatch example, the #10-24 thread is known as "Unified National Coarse," or UNC thread. The #10-32 thread is known as "Unified National Fine," or UNF thread. Metric fasteners have different pitch options as well.

There are some basic rules associated with selecting the length of your fastener. For a threaded fastener, like a bolt or screw, select a fastener with a length that is equal to or slightly longer than the thickness of the parts, the height of the nut, the thickness of any and all washers, plus approximately 2mm (~0.08"). This allows some wiggle room for additional things, like a lock washer, for instance. If a fastener is to be screwed into a threaded hole, make sure that the fastener is long enough to engage an absolute minimum of 3 threads - this is referred to as "3 threads of engagement."

Assembly Basics

One aspect of using fasteners of any kind is attending to the stresses applied by clamping two or more parts together tightly with a nut/screw/washer or similar

Ⓐ Machine Screw Gauge Chart

Gauge	Shank Ø	Gauge	Shank Ø
#0000	0.021"	#4	0.112"
#000	0.034"	#5	0.125"
#00	0.047"	#6	0.138"
#0	0.060"	#8	0.164"
#1	0.073"	#10	0.190"
#2	0.086"	#12	0.216"
#3	0.099"	¼"	0.250"

Staying on Track: Determining Appropriate Rivet Length

This is where grip length comes into play. Samer's time working on experimental and homebuilt aircraft educated him on the selection of rivets, and especially how to select a suitable grip length. Grip length is defined as the length of the fastener's shank that is unthreaded. However, for our purposes, we will extend that to how much of the fastener traverses all the holes. When fastening metal, for example, the rule is to select a rivet whose grip length is at least or slightly more than the combined thickness of all the parts being fastened together. Below is an image of a wing spar with a series of driven rivets used to bring all the parts together.

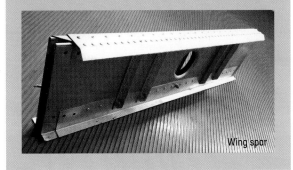

Wing spar

edge, repetitive stresses eventually cause a crack, especially in plastics such as acrylics. Place the center of any hole 1 to 1.5 times the diameter of the fastener's head from the edge. Where possible (if using screws), use a washer to distribute the stresses further away from the hole.

- Avoid drilling oversized holes, as those encourage the fastener to wobble around inside the hole under repetitive stresses and vibration. Eventually, the hole may elongate and the fastener shift, and components may go out of alignment. If a hole is oversized, then use a suitable fastener that fits that hole exactly, or add a bushing to take up the extra space.

- When drilling metal, take the time to "deburr" your holes. Drilling leaves bits of metal sticking up from the edges of the hole. These can prevent fasteners from making full contact with the metal surface, and also prevent your shank from going straight through. You can go out and buy a specialized deburring tool (Figure **A**), but you can also just use a drill bit much larger than the hole you drilled. Figure **B** shows a homemade hole-deburring tool made from a large drill bit bit, covered with rubber air line tubing to make a nice, comfortable handle.

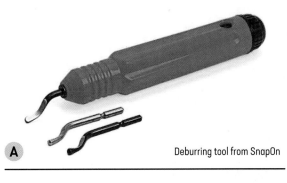

A   Deburring tool from SnapOn

combination. If the holes we put the fasteners through are subject to vibration (if whatever you are making moves at all) and the material itself gets compressed somewhat, it may crack with repeated uses over time. This happens with just about any material, so it's important to consider how many fasteners you use, and also and how far apart they are, since proximity between fasteners is also a source of stress effects on the material.

SPACING FASTENERS

- For very thin or soft sheet materials, such as acrylic sheet and aluminum sheet, space the holes by at least 1.5 times the diameter of the fastener's head. With softer materials such as plastics, space them by 3 times the diameter of the head.

- Use the same guidelines for the edge of the material. If you drill a hole too close to the lose to a material's

B   Homemade hole deburring tool

When tightening a fastener, consider the amount of torque you apply. Although it's unlikely you can apply enough torque manually to shear the shank of your fastener, it bears mentioning that you should not overdo it. This is a judgment call for ad-hoc building since most makers will not have torque wrenches handy while assembling projects. Be aware of it, nonetheless. If you are building something that specifies torque values, buy and use a torque wrench in order to strictly adhere to those values.

Use lock washers and thread lockers (such as from LocTite™) to help prevent fasteners from working loose due to vibration.

Be mindful of drilling holes that may intersect each other. If you are creating, say, a solid rectangular corner bracket, and holes are coming coming in from two or more sides, take care to keep the holes separate, while still obeying the hole spacing guidelines.

When assembling, make sure to allow enough space to to insert your fastener. There has been no shortage of times we have been 90% of the way through assembling something, only to realize that we did not think the sequence through, and found ourselves unable to add that last bolt, screw or rivet because there was no room to use the wrench or other tool required to complete the job.

Keep sufficient clearance between surfaces and your fasteners to ensure nothing is pressing or rubbing against the fastener. Sometimes parts will fail due to repetitive wear (particularly in high vibration conditions), such as cable clamps rubbing against a piece of tubing.

Adobe stock - Jozef Jankola

Corrosion

Makers commonly work with iron, steel, aluminum, copper, or alloys of these metals. All parts made of these metals, including fasteners, will oxidize when they come into contact with moisture. In the case of iron or carbon steel, this oxidation is known as good old-fashioned rust.

If your structure will be exposed to the elements, protect each material from corrosion, as well as the entire assembly. You would think that corrosion only impacts metal parts. But if you are using metal fasteners to assemble non-metal parts, you may also need to address corrosion effects. Even if your project is not made of metal but is secured with metal fasteners, make sure those fasteners are somehow protected from the environment. Here are a few tips:

Some fasteners are sold coated with a corrosion-inhibiting agent (e.g. deck screws are typically painted with a corrosion-inhibiting paint). Galvanized fasteners are steel fasteners that have been dipped into a non-oxidizing molten metal, usually zinc. Other fasteners are plated (typically achieved via electroplating techniques), or coated with substances like nickel or ceramic to prevent corrosion.

When using wood, make sure the wood is sealed, as moisture in the wood and surrounding environment will rust the fasteners. If your project is going to be outdoors or in a humid environment, consider sealing the holes prior to inserting your fasteners, or pour in a sealant as you screw in your fastener. Also, as a final method of protection, seal the entire assembly with primer and paint.

If possible, completely encase the screw or fastener head, particularly if the assembly is permanent and meant to never be disassembled for maintenance. For example, layer a small amount of fiberglass over composite assemblies to cover the screws or bolts, exposing only the shanks if disassembly is required.

Staying on Track: Galvanic Corrosion

When two dissimilar metals are in contact and in the presence of an electrolyte (i.e., a liquid that can conduct electricity), they form a battery, and galvanic corrosion may occur (shown below). One metal forms the anode (negative terminal) of the battery, while the other, dissimilar metal forms the cathode (positive terminal). The electrochemical reaction between the anode and cathode causes the anode to dissolve in the electrolyte. This process causes the anode to corrode much faster than the metals would without being in contact with each other and an electrolyte.

A good example of galvanic corrosion can happen on your car battery's contact poles and master connectors. Since water can be an electrolyte, and water is in the air, the dissimilar metals that make up the car battery and are clamped together will corrode in the presence of the water. Over time, a greenish-white copper oxidation residue will present itself on the poles and connectors, all due to galvanic corrosion.

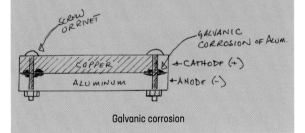

Galvanic corrosion

Fixing Fastener-related Mistakes

We all make mistakes. If you are like us, you will make your fair share — and they will always seem to happen when you are almost done.

We can classify the sort of mistakes we typically make (and their solutions) into a few broad categories:

1. **Poor measurement on our part:** This is not because we don't know how to use a ruler or tape or whatever tool we have, but rather because we make a calculation error. Frequently, that leads to an oblong hole (because it had to be widened to get the fastener's shank in straight) or a second hole drilled very, very close to the first, resulting in a potential weakness at the clamping point.

If this happens, make the hole wide enough to take the next larger size fastener. Or try inserting a spacer (which might require widening the hole even more). Another option is to insert an "eccentric" spacer where the hole in the spacer is not centered in it, so it gives enough deviation to get the bolt in and still have material in the hole. If vibration and strain are unlikely to impact impact the part, add a few washers on the exposed surfaces; toothed washers that will grip the material are preferred.

2. **The Oops Factor:** This is where we momentarily lose focus enough to let the drill slip, or the part under the drill slip, or some other calamity happen that could have been avoided by not rushing or being distracted. Unfortunately, this happens more often than most makers will admit. When we encounter this, we do the same thing as #1 above. It's the same problem after all, but the cause is different.

3. **Improper estimation of material strength:** How many times have you clamped materials down with a nut-and-bolt combination, only to discover the materials cracking and crumbling around the bolt? Usually, this indicates that the material did not have sufficient "meat" to distribute the stresses. This is frequently caused by drilling a hole too close to an edge, or drilling a pilot hole that is too small. (A pilot hole is drilled using a much smaller drill bit than the final hole size, and it is later widened with a larger bit to the final size. Drilling a pilot hole lowers stress on the material being drilled, versus drilling one large hole all at once.) In such cases, there is very little you can do to fix the situation. If you are working in wood or composites, try to epoxy the material together and clamp the parts differently. Otherwise, the part may be ruined and will have to be refabricated.

Staying on Track: "Patching" Mistakes

While fixing cracks around holes in sheet metal, wood, and composites during his airplane builds, Samer has learned about using a "patch." He cuts a small patch of material (same thickness or thicker) and marks holes around the perimeter, far enough from the edge to avoid cracking. The patch has to be large enough to both cover the damage and allow for new holes. In sheet metal, pop rivets are best used to secure the patches. With wood and composites, the patch should be epoxied or screwed over the problem area. Typically, the patch should be of the same material and thickness, but sometimes really large and glaring problem holes require much stronger or thicker materials. With fiberglass, "glass" the area and fill the hole with a gap filler (like "micro-balloons"), then resume your fastening, taking care to account for the change in the makeup of the patched area.

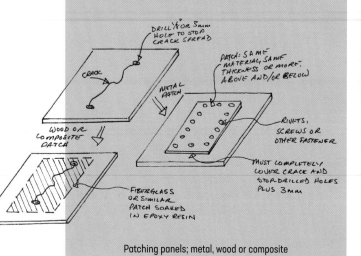

DRILL ¼" OR 3mm HOLE TO STOP CRACK SPREAD

CRACK

PATCH: SAME MATERIAL SAME THICKNESS OR MORE. ABOVE AND/OR BELOW

METAL PATCH

RIVETS, SCREWS OR OTHER FASTENER

WOOD OR COMPOSITE PATCH

MUST COMPLETELY COVER CRACK AND STOP-DRILLED HOLES PLUS 3mm

FIBERGLASS OR SIMILAR PATCH SOAKED IN EPOXY RESIN

Patching panels; metal, wood or composite

An Example of a Fastening System

There are a variety of systems that provide a uniform method for fastening parts together. These are systems that interlock or interact together in some fashion by "forcing" you to adhere to a specific set of materials and parts. It is helpful to use these systems if you seek to create a structure of uniform materials and stick to off-the-shelf parts.

The fastening system we use the most is aluminum T-slot extrusion. (There are several vendors of this system in various incarnations, one of which is called "80/20".) T-slot fasteners (T-slot nuts, for example), brackets, and other system-specific parts (Figure ) are designed to fit in slots in the aluminum extrusion, ensuring that everything lines up. All the pieces are modular, and best of all, the whole assembly can be disassembled when no longer needed and the parts (including the fasteners) can be reused.

A Sample T-slot pwarts

When choosing a fastening system such as this, be mindful of part compatibility. Most systems use proprietary parts, so many parts across differing systems are not always compatible or interchangeable. It is very frustrating to try to get incompatible components to come together. It's like trying to merge multiple jigsaw puzzles into one. Figure **B** shows an example of a flight simulator built using this system.

B Flight simulator built using T-slot components

Adhesives

It can be a bit overwhelming trying to figure out what adhesive to use when bonding materials together. This is where *Make:'s* handy Adhesives Chart, originally published on makezine.com (on page 92), comes to the rescue. The following discussion covers the eleven different adhesives presented in the Adhesives Chart.

- **All-purpose Glue (A):** This type of adhesive can be used to bond a wide variety of materials. It is a good, general purpose glue, and it is typically formulated to dry quickly and to be non-toxic.

- **Fabric Glue (F):** This is typically a non-toxic, urethane-based adhesive. It is designed both to glue fabric to other fabric, and to bond fabric to other materials. Use fabric glue to adhere beads or sequins to a shirt, for example.

- **Spray Adhesive (Sp):** This type of adhesive comes in an aerosol can and is sprayed onto a surface in the manner of spray paint. Depending on the application and formulation of the spray adhesive you are using, items can even be pulled apart and restuck, for example, by peeling off and repositioning a repositioning a picture in a scrapbook after it has been initially placed and adhered.

- **Hot Glue (H):** This is a thermoplastic adhesive that typically comes in the form of solid cylindrical sticks. The solid cylinder of glue is extruded through a hot section of a hot glue gun as a liquid that can be applied to the areas to be bonded. It then re-solidifies as it cooling. This type of glue can be used to adhere a wide variety of materials.

- **Contact Adhesives (C):** This is typically a neoprene rubber adhesive. It dries quickly and provides for a flexible and permanent bond. It is very good for bonding non-porous materials that other adhesives cannot. Unlike most adhesives, you coat the two materials separately with contact adhesive and let them both dry before joining the parts together. But beware: When you touch the two materials coated in contact adhesive together, they bond almost instantly. So, be sure to line up your parts precisely prior to touching them together.

- **Construction Adhesive (L):** This is a general-purpose adhesive used for attaching construction materials, fixtures, and molding to ceilings, floors, and walls. It is typically dispensed using a caulking gun.

- **Ceramic Glue (Ce):** This glue is known as PVA (polyvinyl acetate) glue. It is very good at bonding porous surfaces, such as the edges of broken ceramics, but is not a good glue for bonding non-porous surfaces, such as glazed surfaces of pottery.

- **Silicone (Si):** Silicone adhesive makes for an excellent water-resistant, flexible bond. But as can be seen in the adhesives chart, its uses as an adhesive are very limited. Silicone is more typically used to make joints watertight, rather than actually gluing materials together.

- **Wood Glue (W):** This is a really tough adhesive that is excellent for bonding porous materials (such as wood) together. This not only adheres all types of wood, but also materials such as styrofoam and fabric. Wood glue does tend to be brittle, so do not use wood glue if the resulting joint will be flexed. Wood glue has an interesting property of being water soluble prior to drying, and waterproof after drying.

- **Cyanoacrylate (Ca):** This is more widely known as "superglue." It is a very fast-acting adhesive used for industrial, home, and even medical applications. Cyanoacrylate can be used to bond a lot of material types, both porous and non-porous, including skin! Funny story: Brian had a friend who picked his nose with his finger while using cyanoacrylate to build a model. Let's just say that there was INSTANT bonding... you can surmise the rest of the story.

- **Two-Component Adhesive (2K):** This is more commonly called epoxy. Epoxies consist of a resin and a hardener. When the resin and hardener are mixed together, the adhesive begins to harden. There are many formulations of epoxy with various drying times. They can be used to bond a variety of materials that are typically not easy to bond in any other way.

Adhesives Chart (by *Make:* magazine)

MATERIALS	Paper	Fabric	Felt	Leather	Rubber	Foam	Styrofoam	Plastic	Metal	Ceramic	Glass	Balsa	Cork	Wood
Wood	W	C/W	Sp/C	W/C/Ca	C/Ca	C	2K/H	L/C	2K/C/L	C/Ca	C/Ca	W	W	L/W
Cork	H/W	H/L	W	Ca/C	Ca/C	2K	W	L/Ca	C/Ca	L/Ca	Si	W	W	
Balsa	W	H/W	W	Ca/C	C/Ca	C	2K/H	L/Ca	2K/Ca	L/Ca	C/Ca	W		
Glass	A/W	A	A	A/Ca	Ca	Sp	2K/Sp	C/L	2K/C	2K/C/L	2K/L			
Ceramic	A/H	Ca/A	Ca/A	Ca/A/C	C/Ca	A	Ce/C	L/Ca/C	2K/C/L	Ce/Ca				
Metal	A/H	A	C	C/Ca	C/Ca	C	2K/H	2K/C	2K/C					
Plastic	H/Sp	Sp/C	Sp/C	Sp/Ca	C/Ca	Ca	Ca/C	L/Ca/2K						
Styrofoam	Sp/C	A/H	Sp	A	L	L/A	A/Sp							
Foam	Sp	Sp	Sp	C	C	Sp								
Rubber	Ca/C	A/C	C	Ca	Ca									
Leather	F/Sp	F	2K	C/F										
Felt	A/H	F/H	H/F											
Fabric	A/H	F/H												
Paper	A/W													

Staying on Track: To Cement or Not Cement PVC Pipe and Fittings

As with nearly all PVC projects in this book, you can choose to cement the joints together using PVC pipe cement or not. Just be aware that if you do use cement, the assemblies will be permanently bonded. For somewhat experimental projects like most of the ones you will see throughout this book, we chose to not cement anything together. It makes things really easy to reconfigure, and in some cases to reuse the fittings and pipe segments. However, once you have fully tested a design and are happy with it, feel free to cement your creation together and make it permanent. Also, from a safety standpoint, if the PVC project that you are making requires pressurized air or other fluid in a contained or "fixed volume" system, you definitely need to cement the joints together. To make a stronger bond, we also recommend using PVC primer on each joint prior to applying the cemen Brian's pneumatic cannon, previously mentioned in Chapter 2 shown in the photo below, with his son beside it for a size reference), is a great example of a project that absolutely has to be properly cemented together. Brian has pressurized it upwards to 110 psi!

Pneumatic (air) cannon

Ⓐ Bolted connection between frame and track assembly

Fasteners Used in the Tot-Size Tank

Now let's go into detail about the fasteners used in Brian's Tot-Size Tank. This should give you a a clear, real-world example project example of the use of different fasteners and how to choose between them for a particular area.

Figure Ⓐ shows one of the connection points where the front of one of the track assemblies is attached to the main frame. In this application, Brian chose to use hex bolts with nuts and washers, because the two bolts on the front and the two bolts on the rear of the track were required to support the entire weight of the main frame carrying the drive assembly, batteries, electronics, drive controls, drivetrain, body, and driver. The fasteners in this area needed to be very secure and able to carry a fair amount of load. Since he was fastening the track backbone to a wooden cross member, he could have simply used wood or lag screws. However, Brian did not want to rely on only the small wood area in which the threads would be engaged. So, he chose to use a bolt through both the entire thickness of the track backbone and the frame cross member. Another consideration

A Control handle connection to pivot bracket

for selecting this method was that the tracks would periodically need to be removed from the frame for maintenance and future upgrades. Threading screws in and out of wood is not a good idea, as the wooden threads generated by the screw will wear out and strip fairly quickly.

Figure **A** illustrates another use of a hex-head bolt, nut, and washer to secure one of the control handles. In this case, the bolt is doing double duty. The bolt serves to keep the control handle in place, while also allowing the handle to rotate about it. Effectively, this bolt is also being used as an axle. You may be wondering how, with only one ordinary nut, the handle is able to rotate freely while simultaneously being held rigidly parallel to the control handles mounting bracket. There is a bit of a trick here that is not immediately apparent in the picture. The control handle (on the left) is mounted to a threaded bracket (on the right). The hex-head bolt passes through a clearance hole in the handle and then is threaded into the mounting bracket. On the side of the mounting bracket opposite the handle, a hex-nut is then threaded onto the hex-head bolt (Figure **B**). This hex-nut is then jammed against the fixed, L-shaped bracket, creating a locking effect. This keeps the bolt from rotating while also keeping the bolt axially rigid. To keep the hex-bolt from locking the control handle down rotationally, Brian could have used a nyloc nut. However, this would have allowed the bolt to move around in the mounting bracket, making it a poor axle for the control handle.

Figure **C** shows a close-up view of an edge of the tank's seat (blue) and seat frame (black). To fasten the seat surface to the seat frame, Brian had to use a fastener that would not protrude above the seat surface. He wanted the fastener, when fully engaged, to be flush or very slightly below the seat surface. Considering the seat and seat frame is made from wood, he quite naturally chose a wood screw with a countersink head. This head has a flat top surface with an angled portion below the top, angling down to the shank of the screw. The screw head seats into a countersink hole when the screw is tightened. The depth, angle, and top diameter of the countersunk hole were sized such that the screw's head was flush when it is fully tightened.

In some cases, if the material you are fastening is soft enough, you can simply drive this type of screw into it

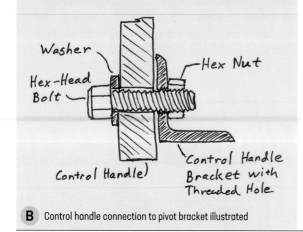

Washer

Hex-Head Bolt

Hex Nut

Control Handle

Control Handle Bracket with Threaded Hole

B Control handle connection to pivot bracket illustrated

C Countersink wood screw fastening seat to seat frame

D Countersink wood screw fastening body frame

E Torx-head deck screw attaching body tube to end of body frame member

Button head wood screw fastening dash to body frame

F

until the head is flush with the surface. The screw head itself effectively machines its own countersink. This is frequently done with wood, as in deck boards. If the deck board is to be fastened with wood screws, simply drive the screw into the deck board's surface, so that it bores its way through (See sidebar *Tracking Further: Screw and Bolt Hole Types* on page 85).

For attaching the seat with countersink wood screws, Brian drilled a tapered countersink hole into the seat, as well as a clearance hole completely through the seat and slightly into the seat frame. Drilling a clearance hole through the seat allowed the thread to engage with the frame and not the seat, enabling the seat to be pulled tight against the frame.

Figure **D** is a close-up of one of the wooden cross-supports of the body, showing another example of a countersink wood screw being used to fasten an area of the tank together. This screw ties the cross-support to one of the metal body frame tubes. With the wooden cross-member in place relative to the metal body frame tube, Brian drilled a vertical hole completely through both pieces. He wanted a tight fit, so this hole was sized such that the diameter was just about equal to the root diameter of the wood screw's thread, forcing the screw to tap its way through the wood of the cross-member and

through the thin metal walls of the frame tube. When fully tightened, this created a very tight, secure joint with only a single screw.

Figure **E** illustrates a situation where Brian used a different type of screw because it was long enough for the application (and he happened to have a bunch on hand). As a maker you may find that it can be challenging, and sometimes even fun, to try to use what you have laying around versus always buying the exact item you need. Here, Brian used Torx-head deck screws that were lying around from another project. He needed a fastener to secure a body frame tube to the end of one of the wooden body cross-members, and one longer than the one used to secure the frame tube, since he had to screw in to the end grain of the wood. Also, he needed a countersink head screw to make the head flush with the outside of the tube when it was fully tightened. The 2½" Torx-head deck screws he already had on hand worked perfectly for this application.

Figure **F** shows one corner of the tank's dash. Considering how visible the dash panel is, Brian wanted to use an attractive screw to fasten it to the body structure. For this application, he used button-head wood screws. This type of screw has a nice-looking rounded head, which is designed to be above the fastened material rather than

A Lag screw used to attach drive axle pillow block bearing to tank's frame

B Lag screw used to fasten control handle pivot bracket to tank's frame

being flush with the surface. Button-head screws also come in machine-screw-thread varieties. So, if you have a fastening task that requires a nice look, consider using button-head screws.

This next fastener example has much more aggressive threads than the previous ones. Figure **A** shows one of the main drive axle pillow block bearings. This particular bearing is mounted to one of the main 2×4 wood frame rails using two ⅜" diameter lag screws. Lag screws are a great choice for fastening something to a relatively soft material, such as wood, when a fairly large fastening force is required. The aggressive, deep threads of a lag screw engage a much larger volume of the wood than a typical wood screw. A lag screw has a much greater difference between the outer and root diameter of its threads. Because of this, you will almost always need to drill a pilot hole for the lag screw so that you do not split the wood.

Figure **B** shows one of the Tot-Sized Tank's control arm brackets. This is another place where lag screws were used. Note that lag screws come in two varieties: fully-threaded, and partially-threaded. In the control arm bracket, Brian used a partially-threaded lag screw. On the drive shaft bearing block (shown in the previous example), Brian used a fully-threaded lag screw to secure the pillow block bearing through its relatively thin flange. Had he used a partially-threaded screw, a good bit of the unthreaded portion of the screw would have been in the frame rail. This would have reduced the number of threads engaged with the frame rail, thereby reducing the possible holding force the screw could generate before the threads stripped out of the wood.

Our final Tot-Size Tank fastener example (Figure **C**) shows one of the points where the body mounts to the frame. The fastener used here is known as a hanger bolt. Brian personally finds these to be really cool: business on one end (machine screw threads) and party on the other end (coarse lag screw threads) – kinda like a mullet hairstyle! It is basically a hybrid fastener. One end has an aggressive, lag-type wood screw thread while the other end has a finer machine screw thread. With this fastener you can basically create a stud embedded in a piece of wood that you can then fasten things to, using a standard machine screw hex nut and washers. To use a hanger bolt, you just drill an appropriately-sized pilot hole and screw the bolt into it. There are a couple of tricks that Brian uses grabbing these, in order to apply enough torque to screw them into the wood. The most obvious way to accomplish this is to simply grab the smooth shank in the middle of the bolt with a pair of vice-grip pliers and rotate. However, there are occasions when you may not have enough room to swing vice-grips a full 360°. In these cases, wouldn't it be nice if you could use a socket wrench to drive the hanger bolt into your piece of wood? Well, there is a solution: jam nuts! Just take two nuts and thread them on to the machine screw end. Using two wrenches rotate one of the nuts clockwise while simultaneously rotating the other nut counter-clockwise jamming them together. If you jam them together with enough force they will be locked torsionally relative to the bolt. Now you can engage the outermost nut with a socket wrench to screw the hanger bolt into the wood. If you don't have jam nuts, use an acorn or dome nut, instead.

We hope that this chapter has served to introduce you to some new fasteners, and has given you some guidance on how to choose and use a variety of common fasteners. There are so many different fasteners, adhesives, and other methods of "sticking" things together, that we would never be able to detail them all here. But this should get you started. As you progress as a maker, it becomes second-nature for you to know what fasteners and adhesives to use in your specific applications. You may even find that over time, you devise ways to create your own "fasten-ating" methods of combining materials!

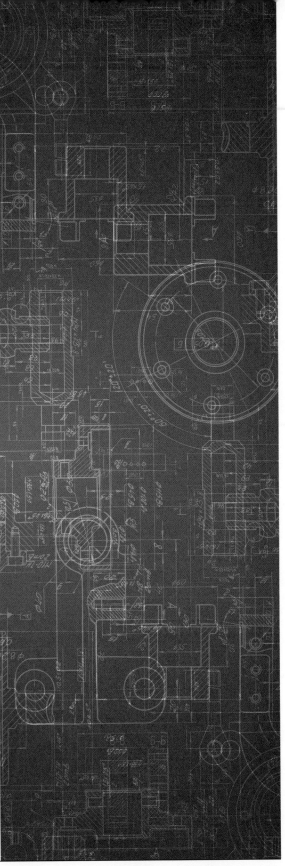

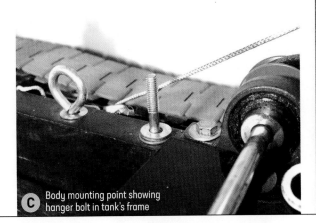

(C) Body mounting point showing hanger bolt in tank's frame

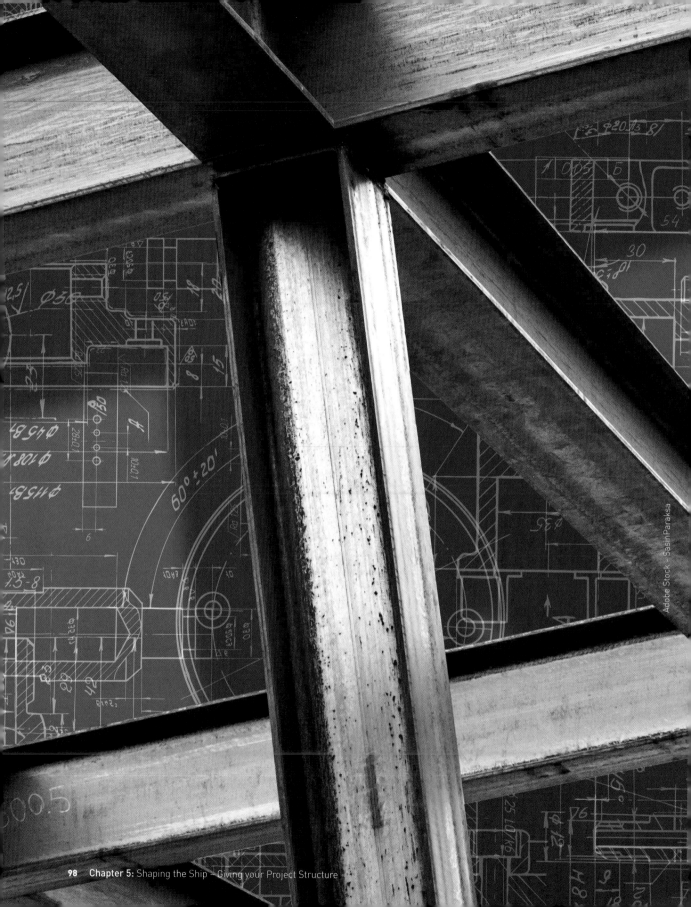

Adobe Stock - SasinParaksa

5

Shaping the Ship – Giving Your Project Structure

For our purposes, we define a **structure** as a conglomeration of materials fastened together in such a way as to form a base or support for a project. As we discuss in Chapter 1, we divide structures into two main categories: static structures and dynamic structures (or systems).

Static structures are structures that have no motion relative to a reference point. In other words, static structures do not move: There are no mechanical components that rotate, or "translate" (move from point to point) in a linear fashion. Static structures are typically used as a platform, foundation, or frame for other components, such as the lower base frame of a typical workshop stool (Figure **A**).

Dynamic structures, on the other hand, have parts that are intended to move when in operation. The tail structure of an RC airplane is an example of a dynamic structure. It includes a servo-actuated rudder and elevator, with a connected tail-wheel that is allowed to rotate with the rudder (Figure **B**). All of these components are designed to work and move together, fixed to the static main frame or fuselage of the airplane.

A Base frame of a typical workshop stool

B Tail structure of an RC airplane

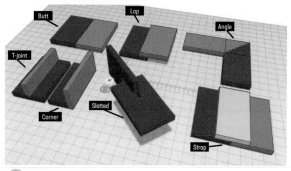

Dynamic structures are always supported by a static structure in the form of a frame, base, or foundation. Figure Ⓐ shows an example of both types of structures being used together. It is a (static) wooden tank frame made of 2×4s and plywood supporting a (dynamic) drive system consisting of an axle, sprockets, and gears.

Design Considerations for Structures

It may seem obvious, but the starting point for determining the design of a structure begins with the same questions that were posed in the project design process: "What are you making? What will it need to do?" Answering these basic questions very quickly leads to defining the general specifications of the project, and these project specifications directly affect the design of the structure. Will the project need to move around, or will it be stationary? Will it need to look nice, be lightweight, be transportable? Will it be exposed to the elements? For example, if your project needs to be mobile, then it may also need to be lightweight. If your project will be exposed to and affected by moisture, it may need to be in an enclosure for protection. If your project will require batteries or bulbs, you will need a way to easily access these areas for battery/bulb replacement. These are all aspects that need to be considered when designing your structure.

Ⓐ Wooden tank frame (static structure) with a drive system (dynamic system)

Ⓑ Some basic joint types

You need to design your project's structural elements to be able to withstand the forces/loads that they will experience during operation. Structures can encounter stress in multiple ways: They may see all forms of stress in an individual and combined way. They may experience bending, tensile, compression, shear, and torsional stresses. Therefore you may find that it is necessary to strengthen your structures to help them resist these forces.

Ⓒ Welded joints using gussets on the inside of the joint

CONNECTING STRUCTURAL MEMBERS TOGETHER - JOINTS

It is very important to consider how to join structural elements/members together when forming a structure. This process is known as **joinery**. Joints, or the meeting and connecting of multiple structural members, can be created in many ways. In fact, the variety of joint configurations and methods can be mind-boggling!

In the realm of wood, there are an amazing number of joint configurations possible! In fact, there are whole

Ⓓ Gusseted corner joint on the outside of the joint

books and classes on wood joinery alone. To give you an idea of the plethora and variety of wood joints, see the "50 Digital Wood Joints Poster" at makezine.com/2014/12/04/50-digital-wood-joints-poster/.

For simplicity's sake, we are limiting our discussion here to a few of the more common joint configurations. Makers find these useful when joining materials such as metal or plastics, but they can also be used on wood (Figure **B**). Many of these joint configurations are illustrated via examples discussed later in this chapter.

COMMON JOINT TYPES

- **Butt Joint:** Two pieces of material in the same plane, pushed together edge-to-edge

- **Lap Joint:** One piece of material placed on top of another, or lapped over another

- **Tee Joint:** One edge of a piece of material placed on the face of another piece of material at a 90° angle, forming a "T" shape

- **Corner (or Angle) Joint:** Usually, two materials with edges cut at 45° angles, and placed to form a 90° angle corner. An angle joint can also be formed when the edge of one material is affixed to the face of another material at its edge

- **Slotted Joint:** Materials containing slots that are fixed together, to interlock

- **Strap Joint:** Two pieces of material in the same plane, pushed together edge-to-edge (i.e., a butt joint) with a piece of material placed on top to span the butt joint, fixing all pieces together

Structural elements or members can be fastened in many different ways: mechanically with brackets, screws, nails, dowels, and other fasteners, or with glues, tapes, and epoxies. Metal joints are often welded, and many joints, especially those carrying large loads, are gusseted, in which the builder adds material where the structural members come together to strengthen the joint.

STRENGTHENING STRUCTURES

One way to strengthen a structure is to add corner brackets or gussets to the joints. Gussets and brackets

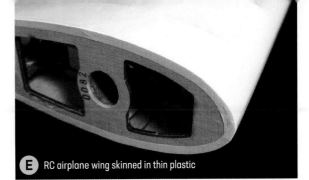

E RC airplane wing skinned in thin plastic

F Welded support-column with cross bracing

help counter the effects of tension, bending, and shear, because the added material prevents motion in a specific direction without adding too much weight. Gusseting can be applied to the inside (Figure **C**) or outside (Figure **D**) of the joint.

Another way to add rigidity and strengthen a structure is to add a "skinning" material to the outside of the structure. Sheet material, such as plywood or sheet metal, can be added to structures as a reinforcing and stiffening material. Securing a piece of sheet material across the outside of a frame helps to eliminate twisting and bending forces. For example, it is common to construct the structure of an RC model airplane wing by covering a lightweight wooden frame/structure with a shrinkwrap film (Figure **E**). This thin, plastic film has a heat-activated adhesive layer on the inside surface. When heat is applied to the outside of the film, the film bonds to the wooden structure and shrinks to form a tight skin.

You can also increase strength and maintain structural rigidity (especially for longer, thinner frame structures) by affixing braces and other members that add stiffness to the structure. This bracing concept is also known as **triangulation**. An extreme (i.e., overkill) example of cross-bracing or triangulation is shown in the welded support column Brian created for a hand-winch crane for his kids' fort (Figure **F**). For more information on triangulation, see *"Tracking Further: Triangulation and Trusses."*

Tracking Further: Triangulation and Trusses

There is a lot in this book about making things that are both strong and lightweight. One method that is frequently used to accomplish this is to build things using frames made of long, thin members connected at the ends. We use various types of braces and other things to add stiffness without too much weight. What these methods of stiffening have in common is that they use triangulation.

To illustrate this concept, refer to the rectangle in Figure ①. Imagine this rectangle represents four bars or members of a frame for your project. The corners of the frame are labeled **A** through **D**. If a force is applied to one of the corners, it is relatively easy to distort the rectangle into a parallelogram, as shown in the second part of the figure. This is because all of the stiffness of this frame relies on how well the joints resist distortion and the bars resist bending. Long, slender bars just aren't very stiff. So in this example, corners **A** and **D** are spread farther apart, and corners **B** and **C** are squeezed closer together.

Now look at the last image in the figure. By adding just one more bar diagonally, connecting opposite corners of the rectangle **C** and **D**, we have turned this frame into two triangles. The interesting thing about a triangular frame is that it cannot distort without forcing one of the bars to change length. Since the new bar **B-C** cannot change length, it keeps the rectangle from distorting. This also means that the joints no longer need to have any resistance to distortion. In fact, it is possible to make a triangular frame with hinged joints that remains rigid!

Engineers refer to these types of structures as trusses. We like trusses because, for design purposes, we only have to worry about forces along the length of each of the truss bars (tension and compression). We can ignore bending within the individual bars, which would require much more complex calculations. If you look carefully, you can find trusses all around you. They're used in bridges, highway signs, large buildings, and possibly in the roof of your house. Here are some photos of common applications for trusses (Figure ②).

You can illustrate triangulation yourself using LEGO bricks or another modular toy, as shown in Figure ③. Make two bars 6 units* (7 studs) long and two more 8

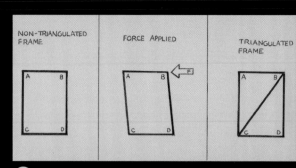

① Simple Illustration of Triangulation

② Three different triangulation truss examples

③ Triangulation example using LEGO bricks

units (9 studs) long (or whatever length you want with the same proportions). Assemble them into a rectangle as shown on the left. Using our previous example, the 6 unit bars are represented by bar **A-B** and bar **C-D**, and the 8 unit bars are represented by bar **A-C** and bar **B-D**. You can see that it is quite flimsy when assembled this way, as shown in the middle image. Now make a fifth bar 10 units (11 studs) long, and place it across two opposite corners **B** and **C** (or **A** and **D**). Now the assembly is quite rigid, as shown on the right. *—John Manly*

*I'm using the term "units" to refer to the number of spaces between the studs on the pieces in the photos, which is one less than the number of studs themselves. So a 6-unit bar has 7 studs, for example.

The next time you pop open a car's hood, look for the bends and raised shapes in the sheet metal. Those areas are called "stiffening beads." Stiffening beads are common in cars and other vehicles, and are used to control twist and bend forces in the sheet metal. Interesting, huh?

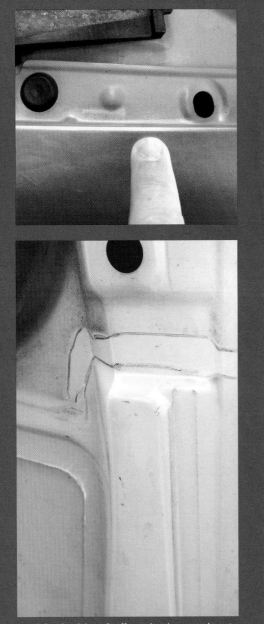

A car's hood with lots of stiffening beads stamped into it

CONSIDERING WEIGHT IN STRUCTURES

There are times when weight is not an issue in a structure. In fact, it is frequently better to start with a simple, if heavy, version to prove the concept for a project, and then remove weight from it in a later, more refined version. In static structures, too much weight is usually not a problem, and attempts to lighten them can introduce extra cost (or time, which is still cost). In fact, some structures actually benefit from a lot of weight. Structures that hold heavy loads, experience intense vibration, or support overhung loads need this extra mass.

But there are definitely times that makers tend to "beef up" or overbuild structures, to ensure strength and durability. It sounds like beefing up a structure is a good idea, but that is not necessarily the case. Overbuilding (i.e., over-engineering) can actually introduce additional cost to your project and unnecessary strain on your structure. In some cases, for instance, where a structure must be mobile or easily transported, you may need to reduce its weight. Here are some ideas to keep your structure light without sacrificing strength:

- If you are adding stiff sheet materials (like plywood or sheet metal) to the outside of your structure, cut holes (called lightening holes) out of the material to reduce the weight without sacrificing strength.

- When working with wood, use adhesives rather than metal fasteners. Properly-applied adhesive joints can be stronger than the material alone.

- If working with metal, consider using blind rivets instead of nuts and bolts. Rivets have a fraction of the weight of nuts and bolts.

- If you need increased strength while also needing to make the structure lighter, consider replacing wood panels with foam-core fiberglass panels.

Tracking Further:
An Example of Joinery —
Flight Simulator Frame

Samer has always wanted to have an in-home simulator for when he isn't able to actually fly. Until recently, his Saitek (flight simulator cockpit instrument) panel was married to his desk along with an old recycled 37" television as the airplane "window." It never felt like the sort of simulator setup that he had experienced elsewhere, but he wasn't shelling out $500 for a pre-made setup that did not meet his lofty specifications. So, he made one for himself!

To illustrate the utility of corner, tee and butt joints, this discussion focuses on the simulator's frame.

Samer's specifications for the simulator frame (Figure Ⓐ) included:

- It had to be built mostly out of material already available. This included several lengths of 20×20mm extrusion and 20×40mm extrusion, a chair, gussets, brackets, screws, T-slot nuts, and various washers from other projects.

- It had to support the Saitek panel, computer, speakers, and all of the electronic and electrical components.

- It had to be on wheels, so it could be moved around with ease.

- It had to accommodate a seated operator weighing up to 400lbs. While Samer doesn't really know anyone who would meet this limitation, he theorized that a 200lb pilot might shift and move around enough to cause the frame to experience higher forces than just the weight of a seated pilot. So he doubled that number as a safety measure.

- It had to support a sliding seat.

The main base of the frame (Figure Ⓑ) successfully supports the entire flight simulator. It accommodates eight wheels, and is able to handle over 400lbs. In addition to a seat, the frame holds the simulator controls, monitors, metal instrument panel, speakers, rudder pedals, a power strip, and a computer (which runs

Ⓐ The flight simulator

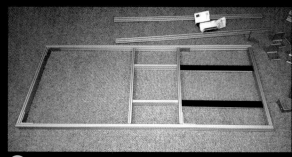

Ⓑ Flight simulator base frame made from T-slot extrusions

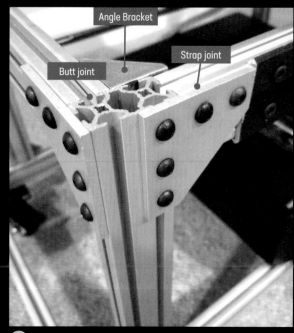

Angle Bracket

Strap joint

Butt joint

Ⓒ Two corner joints on a single vertical member

the simulator software and instruments). The frame provides the foundation on top of which everything else has been built. The selection of extrusions has made it possible to adjust dimensions and correct mistakes by adding more extrusions, or bracing and corner reinforcement, without having to re-engineer the entire design from scratch. The most important decision for this project was the selection of the build material (the extrusions) because they satisfied a number of criteria essential for success:

● They are a forgiving medium on which to build because you can add more supporting pieces, braces, or joint reinforcement

● There is generally no need to drill any holes to build the structure

● They are easily obtainable from multiple vendors online

● They come in multiple dimensional combinations (e.g. 20mm×20mm, 20mm×40mm, and so on)

● They are light, but their shape makes them very rigid

Figure ⓒ illustrates a corner joint in more than one plane. The multiple corner joints corrected a problem that Samer discovered during the build: When an operator pushed or pulled the chair into place, the seat mount deformed under load. Adding a crossmember (and more gusseted corner joints) corrected this issue. Notice the pieces of triangular material on either side of this joint: You can see multiple joints implemented in one location. For example, the triangular pieces are reinforcing a butt joint that effectively combines a butt joint and a strap joint. This is being done on two sides, while on the "inside" of the joint there is an L-shaped angle (or corner) bracket adding to rigidity.

In Figure ⓓ, you will notice a tee joint. The joint connection is made using the outside bracket, as well as an inside L-shaped bracket. There is another L-shaped bracket on the other side of the vertical member that prevents the member from moving.

Sometimes a joint is not sufficient and must be further supported to ensure that any force applied to the far

 A T-joint reinforced at the inside corner

Ⓔ Angled brace to prevent flex in the vertical plane

edges of the members (the pieces joined together) do not overwhelm the joint. To address this, a brace may be necessary. Figure Ⓔ shows just such an example, made of a segment of 20×20mm extrusion, connected using a pair of hinged end (some might even call these butt) joint connectors. The vertical member (discussed above) experiences some flex under load, and this brace prevents the vertical member from leaning inwards towards the operator, while still affording the operator the needed legroom to get in and out of the simulator.

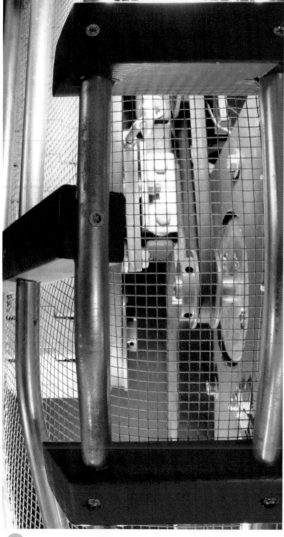

A Tot-Size Tank body

A Static Structure Example – Tot-Size Tank Body

Successfully designed and constructed structures combine function, mechanical integrity, and aesthetics. Brian sought to achieve all of these attributes in designing and building the body for the Tot-Size Tank (Figure Ⓐ).

In general terms, the Tot-Size Tank body is a simple, static structure made up of metal electrical conduit, conduit connectors, 2×2 lumber, plywood, and hardware cloth. Metal electrical conduit is thin-wall tubing designed to be used to route and protect electrical wires. In designing the body, Brian started by asking two obvious yet very important questions: "What must the body achieve? What should the body add to the tank?" The answers to both questions turned out to be quite simple. The body must protect the driver and anyone else around the tank from the moving parts of the drivetrain and control systems, and it needs to add aesthetically and structurally to the tank.

Brian's major criterion for the body structure, from the standpoints of safety, aesthetics and practicality, was that he wanted the operator and others shielded from the internal moving components, while still being able to see the various belts, pulleys, control lines, and sprockets while in operation. He also wanted to be able to observe the inner workings of the tank for easier troubleshooting, were any issues to arise. Using hardware cloth (really coarse metal screen), he achieved the goal of being able to see the mechanics of the tank in action while keeping safety at the forefront of the body's design (Figure Ⓑ).

Additionally, the body had to be stiff. Figure Ⓒ shows all 180lbs of Brian standing on the body. Stiffness and structural integrity are directly linked not only to the materials selected, but also to the geometry by which the materials are configured. Figures Ⓓ and Ⓔ show close-up views of the main structural area of the body. This area of the body exhibits layers of strength through 3D geometry. The top portion of the body transitions from a flat, planar geometry to geometry that slopes down toward the front. Just before the top begins to slope down, you can see a side support that is tied to the upper structure at an angle. The body sides are canted at a slight angle relative to vertical. These various angles of the top and side portions of the body serve to triangulate

Ⓑ Body, showing drive components through hardware cloth

Brian standing on
Tot-Size Tank body

C

D Body, structural integrity through geometry

E

the overall body structure, further increasing its stiffness.

The tubes of both the upper and side areas of the body are terminated into the bulkhead of the body, toward the middle of the tank (Figure **F**). This bulkhead is basically a stiff, vertical wall that serves both to terminate the body tubes, and also as a place to mount the plywood instrument panel. This houses the electrical components such as lights, switches, and various gauges to monitor the electrical system and batteries.

F Body bulkhead, where top and side body tubes are tied in

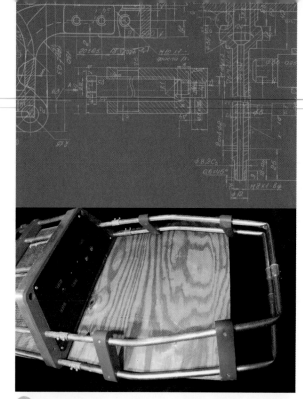

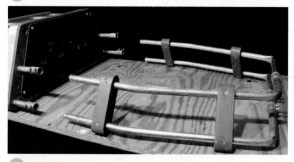

This plywood instrument panel skin also helps to further stiffen the bulkhead (Figure **A**). Lastly, the bulkhead helps keep the driver safe by cutting off access to the moving drive and control parts from the driver area.

When you are thinking through and designing your projects, it is a good idea to try to imagine how you may possibly evolve or repurpose your project in the future This is especially important for larger, more involved projects. Very early in the design of the Tot-Size Tank, Brian started thinking about what to do with the tank when his son outgrew it as a rideable contraption. He honed in on the idea of reconfiguring it to be remotely controlled. To prepare for this future conversion, he decided he could house the future components (i.e., servos and radio system) in the area of the tank currently allocated for the driver. To protect these future components, the body's rear area would need to completely cover this area, which it doesn't right now. So, he designed the body to allow the back half (currently being used) to be easily separated from the front (Figure **B**).

This was achieved by making the rear portion of the body its own structure. The rear body structure was then attached to the front body structure by using four inline conduit connectors. In the future, when it's time to convert the tank to a remotely- controlled vehicle, Brian can simply build a new rear body structure with the same connecting geometry to replace the current one.

There are two distinct advantages to this approach for the body. The first is that Brian only had to build one front structure to protect the driver and people around the tank from the drive and control components. The second is that once he builds the rear enclosed structure for the remotely controlled configuration, he can easily convert it from remotely controlled back to driven from within, by simply changing out the rear body structure.

A Dynamic Structure Example – Tot-Size Tank Tracks

Each track system of the Tot-Size Tank is an example of a dynamic structure. The primary purposes of the track structure are to support the tank and driver, and to provide for a means to power and control the tank. The drive sprocket used to support and drive the track chain is at the front end of the track. The idler sprocket used

A   Tot-Size Tank body showing instrument panel (dash)

B   Rear body section shown separated from front

C   Track structure showing drive belt, pulley and tensioner

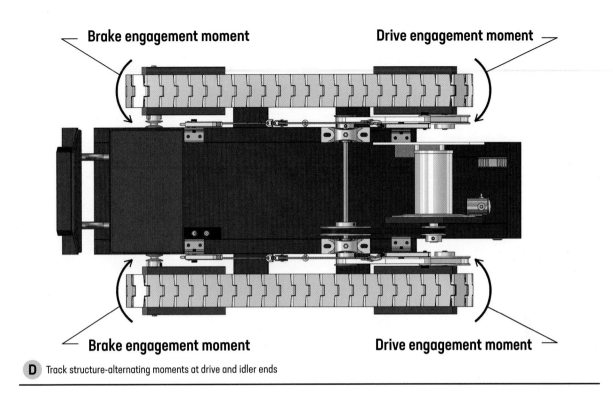

Brake engagement moment

Drive engagement moment

Brake engagement moment

Drive engagement moment

D Track structure-alternating moments at drive and idler ends

to support and tension the track chain is at the other end of the track. Both of these sprockets must be supported by the track structure in such a way that they can rotate freely, but not move in any other **degree of freedom** (see *Tracking Further: Degrees of Freedom and Constraints* on page 110) relative to the structure. In other words, the track structure must keep the center of rotation of the drive and idler sprockets fixed relative to itself and, by extension, the rest of the tank. Considering the fact that the sprockets of the track structure rotate during operation of the tank, it is the first aspect of this structure that makes it a dynamic structure.

There is a second, far less obvious aspect of the track structure that also defines it as a dynamic structure. The interactions of both the drive and brake belts with the track structure (i.e., the changing or dynamic loading of the track structure while the tank is in operation) also make it dynamic.

To send power to one of the tracks, the track's drive belt is tensioned (Figure **C**). The tensioning applies a significant amount of force to the track's drive pulley perpendicular to the drive axle. Considering that the line of action of the belt and drive pulley is off of the centerline of the track,

there is a moment (or rotational force) generated relative to the main frame of the tank that the track structure must be able to take.

When the brake is applied, the same dynamics are at play at the other end of the track structure. Conceptually, the brake works in a similar manner as the drive. When the brake is engaged there is also a moment applied to the track structure. However, the moment at the brake end of the track acts on it rotationally in the opposite direction as that of the drive end. So, the track structure must be able withstand these repeated, alternating moments without flexing significantly, while also remaining static with respect to the main frame of the tank (Figure **D**).

Structures are extremely important to the success of your projects. They need to meet the basic requirements of the projects, and may therefore require methods of strengthening to meet those specifications. In some cases, you may also find you'll need to increase or decrease the weight of a structure, depending on the project parameters. We hope that this chapter serves to guide you in the design and fabrication of the static and dynamic structures that together make up the core of your future projects.

Tracking Further: Degrees of Freedom and Constraints _____

The term "**degrees of freedom**" or DOF is used by engineers to describe how something is free to move in space. To use a simple, two-dimensional example, consider a side view of a car wheel (Figure Ⓐ). The degrees of freedom can be listed as follows:

1. The wheel can move right or left, parallel to the ground.
2. The wheel can move up and down.
3. The wheel can spin around its center.

Any motion of a wheel, or any other two-dimensional object, can be defined as some combination of these three degrees of freedom.

When we are dealing with real systems, they almost always have fewer than the maximum number of degrees of freedom (usually the goal is to have only one). For example, if the wheel is touching the ground, held in contact with it by gravity, it may be moving parallel to the ground and spinning at the same time. We say the vertical motion is "constrained" because the ground won't allow the wheel to move down, and gravity won't allow it to move up unless enough force is applied to overcome it. When something is constrained, the constraint removes at least one degree of freedom (vertical motion, in this case). So does this mean the wheel still has two degrees of freedom? Maybe. If the wheel is on a slippery surface like ice, for example, the horizontal motion and spinning of the wheel can happen independently of each other. In that case, the wheel does have two degrees of freedom (probably undesirable if the wheel happens to be attached to your car). But in normal cases where there is plenty of friction between the wheel and the ground, the two motions of the wheel are constrained together. So the amount of rotation of the wheel is directly related to the distance the wheel moves. This removes the second degree of freedom, leaving only one.

Why do engineers and makers care about degrees of freedom? We know that nearly any useful mechanism has to be properly constrained to function as intended. Most of the time, a mechanism has some sort of input motion or force, and a single output motion or force. This requires a mechanism with only one degree of freedom, as shown in Figure Ⓑ on the left. If it had zero degrees of freedom, it would be a static structure, and not produce any output, as shown in the middle figure. If it had two or more degrees of freedom, it could have multiple outputs for a single input. This usually isn't the intent.

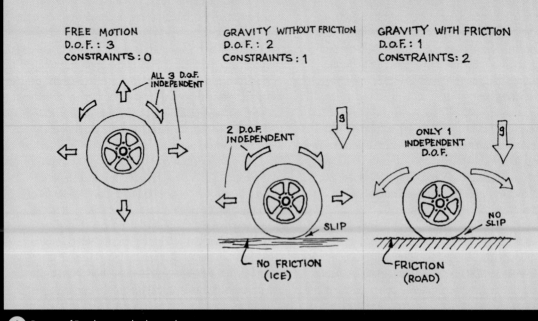

FREE MOTION
D.O.F. : 3
CONSTRAINTS : 0

ALL 3 D.O.F. INDEPENDENT

GRAVITY WITHOUT FRICTION
D.O.F. : 2
CONSTRAINTS : 1

2 D.O.F. INDEPENDENT

SLIP

NO FRICTION
(ICE)

GRAVITY WITH FRICTION
D.O.F. : 1
CONSTRAINTS : 2

ONLY 1 INDEPENDENT D.O.F.

NO SLIP

FRICTION
(ROAD)

Ⓐ Degrees of Freedom-car wheel example

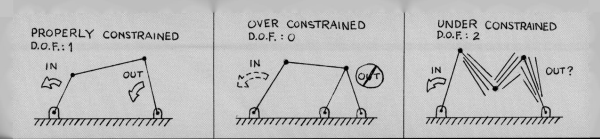

Extending this concept to a three-dimensional object, there are a maximum of six degrees of freedom. To visualize this, consider an airplane in flight (Figure **C**). The degrees of freedom can be listed as follows:

1. The airplane can move forward or backward (for a real airplane, just forward — rotary-wing aircraft, like helicopters can also move backward), along its longitudinal axis, which is an imaginary line along the length of the airplane.
2. The airplane can move side-to-side along its lateral axis, which is an imaginary line parallel to the wings.
3. The airplane can move up and down along its vertical axis. Put another way, it can gain or lose altitude.
4. The airplane can pitch up or down, which is a rotation about its lateral axis.

5. The airplane can roll right or left, which is a rotation about its longitudinal axis.
6. The airplane can yaw right or left, which is a rotation about its vertical axis.

Any movement of an airplane can be defined as movement in one or more of these degrees of freedom. This motion can be quite complex, as anyone who has flown will attest!

There are equations engineers use to calculate the degrees of freedom in complex designs. That's beyond the scope of this book. But engineers and makers alike need to visualize how their designs can move, and add constraints to control how they operate. Like any other skill, it takes practice. —*John Manly*

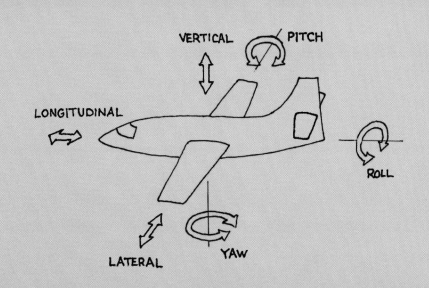

6

Levers – Handling the Suspense

Levers are one of the most common simple machines on earth. When you think of a lever, you probably think of a handle or arm that is attached to and moves about a hinge – and you'd be right. But don't sell these powerful machines short just because they are so simple!

A lever is a simple machine made up of a rigid body with three basic elements; a load, a pivot point (or fulcrum), and an applied effort. These simple machines make life easier for us by amplifying force, amplifying motion, and/or changing direction. A lever can be used alone or in series, depending on the amount and type of work that is needed.

Levers have been valuable machines throughout history. Ancient Egyptians used a lever called a **Shaduf** (or **Shadoof**) to irrigate their fields (Figure **A**). This simple lever on a frame lifted water from its source and allowed it to be poured into the fields. Levers have also been used in more destructive ways. Catapults and trebuchets utilize levers to fling projectiles into and over city or castle walls. Levers even aided the early explorers: They were the tillers and rudders used to steer ships and boats of all kinds.

Interestingly enough, your arm is an example of a lever. Or, to be more precise, it consists of a series of connected levers that allow freedom of movement in more than one direction through ball joints, rather than hinges. This series of levers multiplies the amount of work that can be done. Let's illustrate: Grab a ball and throw it as far as you can. Then, grab another ball and throw it as far as you can, but this time use only your forearm (pretend your upper arm is glued to your side and can't move). The distance for the second throw is far less than the first throw because you only used one lever; you did not get the advantage of the work of more than one lever in a series.

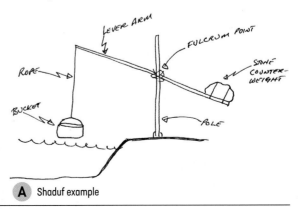

A Shaduf example

A Tank showing control levers

B Seesaw - class 1 lever

C Class 1 lever in tank

In Brian's Tot-Size Tank example (Figure **A**), you can see the use of levers in the control arms. These control arm levers are attached to another lever via cables. The second lever transmits the energy generated by the motion of the first lever, creating tension in the track drive belts.

Samer's Adult-Size Tank also uses levers as a suspension component. He explains the system in detail in *Tracking Further: Adult-Size Tank Suspension Component - Trailing Link* on page 130.

Levers are in such common use today that you probably don't even notice them.   Here are some everyday examples: shovel, claw hammer, brake and gas pedals, stapler and staple remover (two levers), bottle opener, and scissors (two levers). They are everywhere!

Classes of Levers

Levers are not all created equal. They are not all used used in the same ways, and they do not offer the same amount or type of work. Levers are categorized into three "classes," based on the relative position of the load, the fulcrum, and the force required to accomplish work along the rigid body (effort).

A Class 1 lever is defined as a rigid body with the fulcrum between the load (F_L) and effort (F_E). It is commonly illustrated by a playground seesaw or teeter-totter, as shown in Figure **B**. A seesaw's fulcrum is in the center of a rigid board that has seats at each end. The child seated on the side going down represents the effort part of the lever, while the child traveling up represents the load. Referring back to the Tot-Size Tank example, the main control handles of the tank are also Class 1 levers (Figure **C**). Note that the load and effort points do not have to be equidistant from the fulcrum. In the case of the Tot-Size Tank's main control handles, the location that the effort force is applied (where you push the handle with your hand) is significantly further away from the fulcrum than the load from the cable attached to the lower end of this lever. We discuss this system and the reasons for the geometry later in the chapter.

A Class 2 lever has the fulcrum at one end and the effort (F_E) at the other, with the load (F_L) in between the fulcrum and effort. A common example of a Class 2 lever is a wheelbarrow. The wheel/axle represents the fulcrum of the lever, the stuff you are carrying in the wheelbarrow

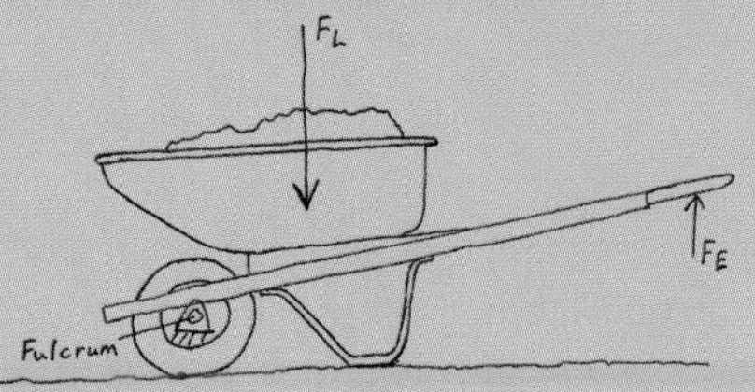
F_L
F_E
Fulcrum

D Wheelbarrow - class 2 lever

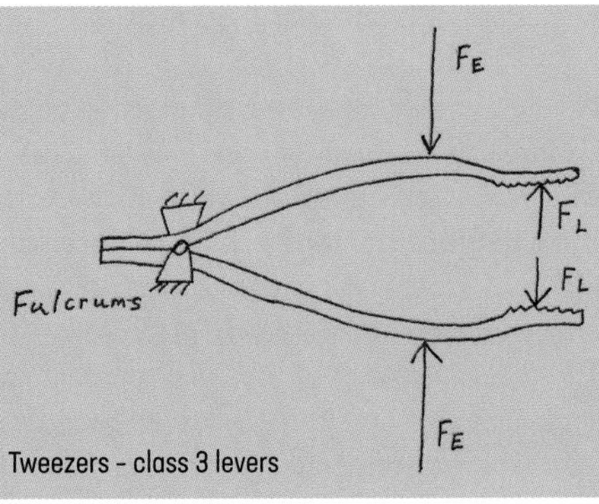
F_E
F_L
F_L
Fulcrums
F_E

E Tweezers - class 3 levers

and the weight of the wheelbarrow not supported (less the wheel) represents the load, and the lifting force applied to the handles represents the effort (Figure **D**).

Given the definitions of Class 1 and Class 2 levers, we bet you can deduce the combination that defines a Class 3 lever...yes, you are correct! The fulcrum is located at one end of a Class 3 lever, the load (F_L) is located at the other end, and the effort (F_E) is between the load and the fulcrum. Tweezers are a good example of a Class 3 lever, or more accurately, a combination of two Class 3 levers working together. The junction of the two sides is the fulcrum, and the effort is the pressure applied to the sides. The two sides of tweezers come together on the opposite end of the fulcrum generating the load force. So, the next time you extract a splinter from your hand, marvel at the work accomplished by a pair of Class 3 levers (Figure **E**)!

Now that you know what defines the three classes of levers, let's consider the type of work each class performs, and why one class of lever would be used for a given situation versus another class. To discuss this topic, we must first introduce a few new terms: mechanical advantage, the "law of the lever," and equilibrium.

Mechanical Advantage (MA) is not specific to just levers: All types of mechanical systems can apply mechanical advantage. In the simplest of terms, mechanical advantage is the ratio of a system's output force to its input force. In other words, it's the amount of work you get out versus the amount of work that you put in.

Tracking Further: Calculating Mechanical Advantage

Calculating mechanical advantage (MA) starts with understanding the idea of equilibrium. In reference t our seesaw or Class 1 lever pictured below, the ***mome*** or ***torque*** caused by the load force (F_L) multiplied by th distance the load force is from the fulcrum (d_L) must equal the effort force (F_E) multiplied by the distance th effort force is from the fulcrum (d_E). When this is true the system is said to be in equilibrium.

If we rearrange the equation representing this equilibrium (Equation 1 in the illustration below) such that the force variables are isolated from the distance variables, we get a new equation (Equation 2), which is the equation for MA. Note that by dividing the load force (F_L) by the effort force (F_E) you can calculate MA. However, you can also calculate MA by dividing the distance the effort force is from the fulcrum (d_E) by th distance the load force is from the fulcrum (d_L).

For our numerical example, we need to calculate the result of the load force (F_L) at a given effort force (F_E). Using the equation for equilibrium (Equation 1), we solve for the load force (F_L), resulting in Equation 3. Once we use Equation 3 to calculate the load force (F_L we can then simply divide it by the effort force (F_E) to g the mechanical advantage (MA). MA is a dimensionle quantity. In other words, the unit of measure for the numerator is the same as the unit of measure for the denominator so they cancel each other out (in/in = dimensionless).

In our example, using an effort force (F_E) of 10 lbs and distance of 12" for between the fulcrum and effort for (d_E) and 4" for between the fulcrum and load force (d_L we have a force-multiplying lever that results in a loa force (F_L) three times greater than the input or effort force (F_E). In other words, MA = 3.

Now, if you want, you can take this even further and derive the load force (F_L) by multiplying MA by effort force (F_E). This means 3×10 lbs, which is 30 lbs.

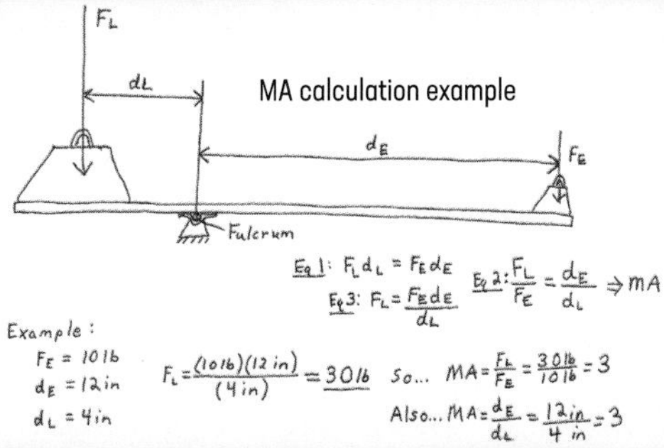
F_L
d_L
d_E
F_E
Fulcrum
Eq 1: F_L d_L = F_E d_E
Eq 2: F_L / F_E = d_E / d_L ⇒ MA
Eq 3: F_L = F_E d_E / d_L
Example:
F_E = 10 lb
d_E = 12 in
d_L = 4 in
F_L = (10 lb)(12 in) / (4 in) = 30 lb
So... MA = F_L / F_E = 30 lb / 10 lb = 3
Also... MA = d_E / d_L = 12 in / 4 in = 3

MA calculation example

A Claw hammer and nail

In the case of levers, mechanical advantage, also termed **leverage**, is the output load force (F_L) exerted by the lever, divided by the input effort force (F_E). In mathematical terms, it is MA=F_L/F_E. See *Tracking Further: Calculating Mechanical Advantage* on page 115 for more information.

The **law of the lever** states that if the effort (or input force, F_E) is further from the fulcrum than the load (or output force, F_L), then there is an amplification of force. Most levers in the form of hand tools are intended to amplify force. Think of a claw hammer removing a nail (Figure **A**). The effort (input force) applied on the handle is further away from the fulcrum (pivot point) on the hammer than the fulcrum is to the nail, giving an amplification of force to the claw and nail (output force). The nail can be pried out of the board by this amplification of force.

The opposite is also true: if the point of effort (input force, F_E) is closer than the load (output force, F_L) to the fulcrum, then the lever results in a reduction of force. Why would you ever want a device that *reduces* the amount of output force relative to the force applied? The answer is that this provides greater velocity (or speed) and greater displacement (or movement)! With a force-reducing lever configuration, the velocity and displacement of the effort end of the lever are less than the velocity and displacement of the load end of the lever because the load end is further away from the fulcrum.

An example of a lever with a reduction of force is a catapult. The catapult is not a tool that is intended to apply a small amount of force to move a heavy load a short distance; It is designed to move a smaller object over a great distance quickly. (For you history and math buffs out there, the law of the lever was proven by Archimedes using simple geometry in the 3rd century.)

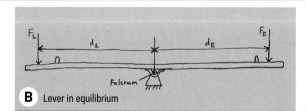

B Lever in equilibrium

For a lever to be in **equilibrium** (or balanced), the load force (F_L) multiplied by the distance from the fulcrum that the load force is applied (d_L) must be equal to the effort force (F_E) multiplied by the distance from the fulcrum that the effort is applied (d_E), or $F_L d_L = F_E d_E$. An example of a system in equilibrium is shown in Figure **B** .

So, considering the law of the lever, dividing the output (or load, F_L) force by the input (or effort, F_E) force returns mechanical advantage as a result. Another way to calculate mechanical advantage is to divide the distance the (input force, F_E) force is applied from the fulcrum by the distance of the output (or load, F_L) force from the fulcrum. F_L/F_E= d_E/d_L = MA describes this approach in mathematical terms. Don't fret; this will all become much clearer when we dive into our first lever project, the "Classy Lever Contraption."

Mechanical advantage can be equal to one, less than one, or greater than one. If a lever has a mechanical advantage of one, then the effort is exactly equal to the load. In other words, if you push down on a lever with 10lbs of force, and the output force also equals 10lbs, then the lever's mechanical advantage equals one. A mechanical advantage equal to one doesn't seem like it would be useful, but even though there is no amplification or reduction in force, there is a change in direction. Pushing down on the effort end causes the load end of the lever to go up. So, in this case, this change in direction is the type of mechanical work that the lever accomplishes, rather than amplification or reduction of force. Remember: According the definition of simple machines, they can cause a change in force, a change in direction, or both. This principle also applies to levers.

A Class 1 lever is the only class of lever that can assume a mechanical advantage of one. It is also the only class of lever that results in a change of direction, although a Class 1 lever can be designed to have a mechanical advantage less than one or greater than one, depending on where the fulcrum is placed relative to the output load and input effort. A Class 2 lever always provides amplification of force, or a mechanical advantage greater than one. The effort force is applied to a Class 2 lever further from the fulcrum than that of the load. The purpose of a lever configuration that yields a Class 3 lever is for increasing speed or displacement of the load end of the lever. Hence, a Class 3 lever always reduces force.

The mechanical advantage for a Class 3 lever is always less than one, because the effort is closer to the fulcrum than the load (Figure C). Note that the three lever levers presented in the figure are balanced: all three are in equilibrium.

Alright! Enough of this theoretical engineering talk. Let's make something!

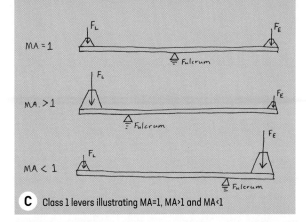

C Class 1 levers illustrating MA=1, MA>1 and MA<1

CLASSY LEVER CONTRAPTION

Our first lever project, called Classy Lever Contraption, shows you how to make a lever representing each of the three classes. These three levers interact with each other as one system that integrates all three lever classes. After building this project, you will have a much better understanding of the three lever classes and what they have to offer for your future projects. You will also pick up a few general making tips and tricks along the way. So, let's get started!

The Classy Lever Contraption is made up of a supporting frame and three levers, each representing the three lever classes. A small chain connects the uppermost lever, the Class 1 (red) lever, to the next lever down, the Class 2 (green) lever. The Class 2 lever is connected, also by a chain, to the lowest lever in the system, the Class 3 (blue) lever (Figure D).

The Class 1's load end and the Class 2's effort point are where the Class 1 and Class 2 levers are connected (Figure D). The Class 2 lever's load point connects to the Class 3 lever's effort point. The connections between the levers serve to create a single system. A system is simply a group of discrete items interacting with each other. The Class 1 lever experiences a single input force on its effort end. This force is then transferred through the system resulting in a single output force

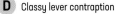

D Classy lever contraption

at the load point of the final, Class 3 lever. The forces involved within the system are what link all three levers into one system. The two chains linking the levers to each other carry the internal system forces. See *Tracking Further: Classy Lever Contraption in Math Form* on page 118 for a more in-depth, mathematical explanation of the Classy Lever Contraption system.

Tracking Further: Classy Lever Contraption in Math Form

The Classy Lever Contraption is a mechanical system comprised of three individual levers, each representing one of the three classes, linked together via small chains. The Class 1 lever is linked from its load force (F_2) to the effort force point on the Class 2 lever (F_3). The load force point of the Class 2 lever (F_4) is linked to the effort force point of the Class 3 lever (F_5) as shown in Figure Ⓐ.

Our mathematical goal in this Tracking Further is to determine how much force is required to lift an item of a specific weight using this lever system. To be more precise, we want to calculate the system's input effort force (F_1) for a given or known system output load force (F_6). In order to determine F_1, we must derive a system equation that combines the equilibrium equations of each of the individual levers with equations that represent the relationships between the levers.

To begin, write out the equilibrium equation for each of the three levers, as expressed in the 3 equations below:

- Class 1 Equilibrium: $F_1 d_1 = F_2 d_2$
- Class 2 Equilibrium: $F_3 d_3 = F_4 d_4$
- Class 3 Equilibrium: $F_5 d_5 = F_6 d_6$

Since these equations represent only individual parts of the overall system, they must be mathematically linked to represent the system. The links are represented by the physical chains connecting the levers. The Class 1 and Class 2 connection is a constant, equal force pulling down on the Class 1 lever and up on the Class 2 lever, at points represented as F_2 and F_3, respectively. The same is true for the link between the Class 2 and Class 3 levers, represented as F_4 and F_5, respectively. These link equations are expressed as:

- Class 1 to Class 2 Link: $F_2 = F_3$
- Class 2 to Class 3 Link: $F_4 = F_5$

Now, utilize the equilibrium equations and link equations to solve for the "F" variables. Use the Class 2 equilibrium equation to solve for F_3. Then, substitute the resulting F_3 equation into the Class 1 to Class 2 link equation to solve for F_2, as shown in the following equations.

- Class 2 Equilibrium: $F_3 d_3 = F_4 d_4$

  Solve for F_3: $F_3 = \dfrac{F_4 d_4}{d_3}$

- Class 1 to Class 2 Link: $F_2 = F_3$

  Combine and Solve for F_2: $F_2 = \dfrac{F_4 d_4}{d_3}$

Repeat the calculations utilizing the Class 3 lever equilibrium equation and the Class 2 to Class 3 link equation to solve for F_4, resulting in the following equation:

- Class 3 Equilibrium: $F_5 d_5 = F_6 d_6$

  Solve for F_5: $F_5 = \dfrac{F_6 d_6}{d_5}$

- Class 2 to Class 3 Link: $F_4 = F_5$

  Combine and solve for F_4: $F_4 = \dfrac{F_6 d_6}{d_5}$

So far, we have identified equations to represent the relationships between the Class 1 and Class 2 levers ($F_2 = \frac{F_4 d_4}{d_3}$) and Class 2 and Class 3 levers ($F_4 = \frac{F_6 d_6}{d_5}$), and have solved for F_2, F_3, F_4 and F_5. (The variable F_6 is a "known" weight, so it is also resolved.) All that remains is to represent the relationship for the system as a whole. This requires solving for F_1, which is our overall goal.

We need to combine the equations that we have already derived to determine F_1. This require a few steps. First, combine the Class 1 and Class 2 relationship equation with the Class 1 equilibrium equation to pull in variables that can be linked to the Class 3 lever. The resulting equation is represented below:

- Class 1 Equilibrium: $F_1 d_1 = F_2 d_2$

- Class 1 to Class 2 Relationship: $F_2 = \dfrac{F_4 d_4}{d_3}$

  Combine and Solve for $F_1 d_1$: $F_1 d_1 = \left(\dfrac{F_4 d_4}{d_3} \right) d_2$

Now, we need to incorporate the F_6 variable, which is found in the Class 2 to Class 3 relationship equation.

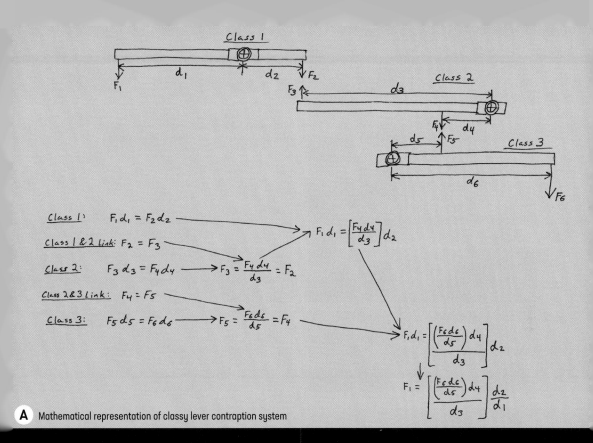

A Mathematical representation of classy lever contraption system

So, combine the Class 2 and Class 3 relationship equation with the equation we just derived for $F_1 d_1$. This brings the entire system relationship together, and the final equation is represented here:

- Class 2 to Class 3 relationship: $F_4 = \dfrac{F_6 d_6}{d_5}$

- Equation solved for $F_1 d_1$: $F_1 d_1 = \left(\dfrac{F_4 d_4}{d_3} \right) d_2$

- Combine: $F_1 d_1 = \left[\dfrac{\left(\dfrac{F_6 d_6}{d_5} \right) d_4}{d_3} \right] d_2$

- Solve for F_1: $F_1 = \left[\dfrac{\left(\dfrac{F_6 d_6}{d_5} \right) d_4}{d_3} \right] \dfrac{d_2}{d_1}$

With this final equation and the known geometry of the levers in the system, you can calculate the amount of force required (F_1) to lift a given weight (F_6) utilizing your Classy Lever Contraption.

Whew! That seemed intense, didn't it? Really, though, each step is quite simple and logical. Use Figure A to walk through the explanation, and it makes more sense.

You may have noticed that this derivation makes some assumptions in its formulation. The weights of each lever arm is not factored in. Also, the weight of the small connecting chains and eyebolts are assumed to be negligible. This is an important principle in engineering. There are many times when the only way you can practically derive an equation that describes a system is by making some educated assumptions. In the case of our Classy Lever Contraption, the weights of the lever arms and that of the chains and eyebolts influences the physical outcome of the system in such a small amount that they can be neglected for simplification.

So there you go! Through the mathematical analysis of the Classy Lever Contraption, you now know how to look at a system of discrete, simple machines as one system

Materials:

» **PVC Pipe – ½" × 5' – standard schedule 40** (see cut list for details), **3 lengths** (MMC* part # 48925K91)

» **PVC Fitting – Elbow (90°) – ½" , qty 4** (MMC part # 4880K21)

» **PVC Fitting – Cross – ½", qty 3** (MMC part # 4880K241)

» **PVC Fitting – Tee – ½", qty 12** (MMC part # 4880K41)

» **PVC Fitting – Side Outlet Elbow – ½", qty 4** (MMC part # 4880K631)

» **8-32 Eyebolt, 1⅛" shank length with nuts, qty 6** (MMC part # 9489T46)

» **#14 Chain** (see cut list for details), **1' length** (MMC part # 3603T76)

» **¾" Oak (Hardwood) Dowel** (see cut list for details), **qty 1' length** (MMC part # 96825K19)

» **6-32 Nyloc Nuts** (optional), **qty 6** (MMC part # 90631A007)

* MMC = McMaster-Carr www.mcmaster.com

Tools Required

» **Saw** (either handsaw, miter saw, or PVC pipe cutting tool)

» **Pliers**

» **Drill**

» **³⁄₁₆" Drill Bit**

» **Rubber Mallet**

» **Fine-Tipped Marker** (such as a Sharpie)

CUT LIST for pvc pipe, dowel, and chain

Qty	Material	Cut Length
2	PVC pipe	14¹⁄₁₆"
2	PVC pipe	10½"
1	PVC pipe	7⅝"
2	PVC pipe	7½"
2	PVC pipe	6½"
4	PVC pipe	5"
2	PVC pipe	4¹⁄₁₆"
6	PVC pipe	4"
3	PVC pipe	3¼"
6	PVC pipe	1½"
3	Hardwood dowel	4"
2	Chain	4 links

CONSTRUCTION

The first step in building the Classy Lever Contraption is to gather all of the required tools and purchased components. The components include PVC pipe, PVC pipe fittings, a wooden dowel, chain, and eyebolts. Referring to the materials list and the cut list, cut all of the PVC pipe lengths from three 5' lengths of ½" standard schedule 40 PVC pipe. We found it immensely helpful to label each piece of cut pipe with its length using a fine-tipped marker such as a Sharpie, as shown in Figure **A**. See *Tracking Further (B3) – PVC Pipe Cutting Techniques* on page 34 for guidance on cutting PVC pipe.

Next, cut the three lengths of wooden dowel that will serve as the axle on which the three levers will pivot. The pivot axles will serve as the fulcrum for the levers.

Lastly, make four identical lengths of chain. Using pliers, bend the link open below the bottommost link you wish to retain. You can then "break" the chain at this link. For our project, make two lengths of chain comprised of four links each. In other words, break the chain by bending the fifth link open and removing the rest of the chain from this point on (Figure **B**).

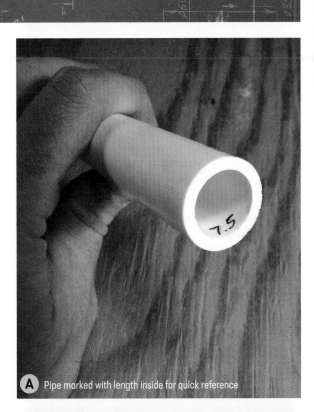

A Pipe marked with length inside for quick reference

B Breaking chain into four link length

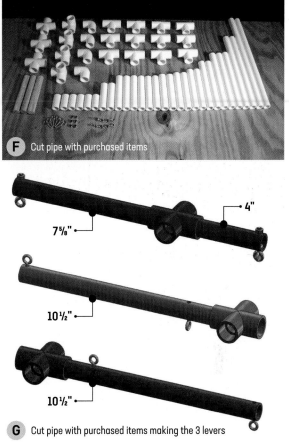

F Cut pipe with purchased items

7⅝"

4"

10½"

10½"

G Cut pipe with purchased items making the 3 levers

All of the items necessary for construction, including the required purchased parts and materials already cut to length, are shown in Figure F. Now, we are ready to start construction (Figure **F**).

The first thing that we are going to do is build the three levers (Figure **G**). Each lever consists of one cross fitting, which serves as our fulcrum, and two lengths of PVC pipe (for the Class 1 lever) or a single length of PVC pipe (for the Class 2 and Class 3 levers). For the Class 1 lever, take a 4" and 7⅝" piece of pipe. Drill a ³⁄₁₆" diameter hole through both walls of both pipes ¼" from one end. Now, assemble the Class 1 lever by seating both lengths of pipe on opposite sides of the cross fitting with the holes you drilled on each end of the lever. Be careful to align the pipe such that the holes are both vertical relative to the cross fitting, as shown in Figure **H**. For tips on aligning PVC pipes and fittings, see *Staying on Track: Pipe Fitting Alignment Made Easy* on page 123.

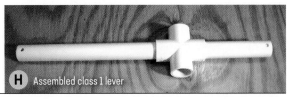

H Assembled class 1 lever

Now we will make both the Class 2 and Class 3 levers at the same time, since they are geometrically identical (Figure **A**). Question to ponder: If the Class 2 and Class 3 levers are the same geometrically, what makes one a Class 2 lever where the other is a Class 3 lever? These two levers each use a 10½" length of PVC pipe. In both pieces of pipe, we will need to drill two ³⁄₁₆" diameter holes. For both pipes, measure ¼" from one end and drill a hole completely through as you did for the Class 1 lever. Now from the other end, measure 2" over and drill a second hole oriented the same as the first hole in both pieces of pipe. Fit the 10½" length of pipe into a cross fitting, making sure that the hole that is 2" from one end is nearest the cross fitting. Also, ensure that the holes are vertical relative to the cross fitting. Repeat to make both the Class 2 and Class 3 levers.

Now we need to make a frame to support the three levers in a proper geometric configuration, so that we can connect them into a single lever system (Figure **B**). The frame will be constructed in three sections to facilitate final assembly.

The first section of the frame is the base (Figures **C** and **D**). Using the components shown in Figure B, assemble the base, ensuring that all four side outlet elbows are oriented with their open sockets facing up.

The next section of the frame to be built is the front, vertical section (Figures **E** and **F**). Again, use the image in Figure B for assembly details. It is critical that the tee-fittings and elbows are oriented in the correct direction, as shown in the diagram.

The third and final section of the frame is the rear, vertical section (Figure **G**). This section also includes three additional pieces of pipe that serve as supports between this section and the front, vertical section. Again, ensure that the fittings and support pipes are oriented properly, as shown in Figure **H** .

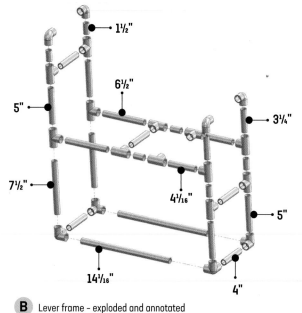

1½"

6½"

5"

3¼"

7½"

4¹⁄₁₆"

5"

14¹⁄₁₆"

4"

B Lever frame – exploded and annotated

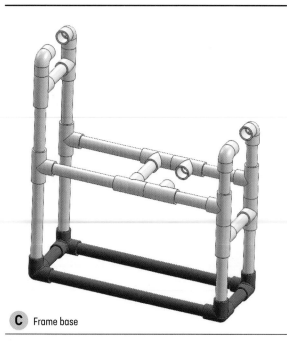

C Frame base

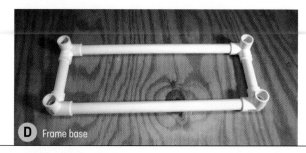

D Frame base

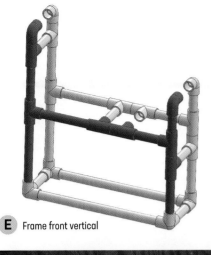

E Frame front vertical

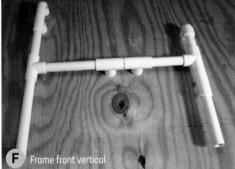

F Frame front vertical

G Frame rear vertical

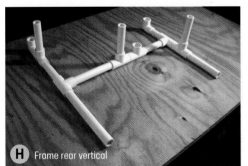

H Frame rear vertical

Staying on Track:
Pipe Fitting Alignment Made Easy

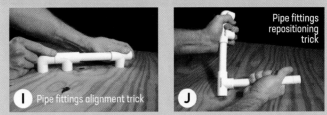

I Pipe fittings alignment trick

J Pipe fittings repositioning trick

A simple and effective way to align pipe fittings is to adjust them as much as possible with your hands (Figure **I**). Before pounding them together with a mallet, lay them on a flat surface. Twist the fittings until they are all flush with the surface (or at a 90° angle, if that is what is needed). Once satisfied, stand the pipe vertically, and carefully hit the assembly with a mallet until everything is fully seated.

It never fails that at least one fitting gets seated without proper alignment, and once they are seated, fittings and pipes are quite difficult to twist! An easy method of realigning/repositioning pipe fittings once they are seated onto pipes is to use an extra piece of pipe as a handle (Figure **J**). Insert the extra piece of pipe into the fitting opening that is at a 90° angle to the attached pipe, but do not completely secure the pipe by seating it into the fitting - this is meant to be temporary. Using the two pipes as handles, twist the pipes away from each other until the fitting is properly aligned, or (in case you inserted the wrong length of pipe) until the fitting comes off of the pipe. Did you notice that this extra pipe is also a lever?

Also, there will be times that you will need to separate the pipes and fittings. If the extra pipe handle method described above doesn't work, there is another way. Insert a dowel through the open end of the pipe and into the fitting, allowing it to press on the end of the pipe inside the fitting. When everything is in place, hold the pipe firmly and use your mallet to hit the dowel against the fitting, dislodging it from the pipe (Figure **K**).

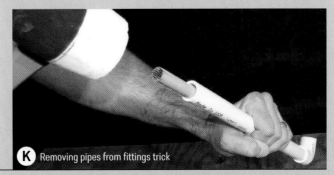

K Removing pipes from fittings trick

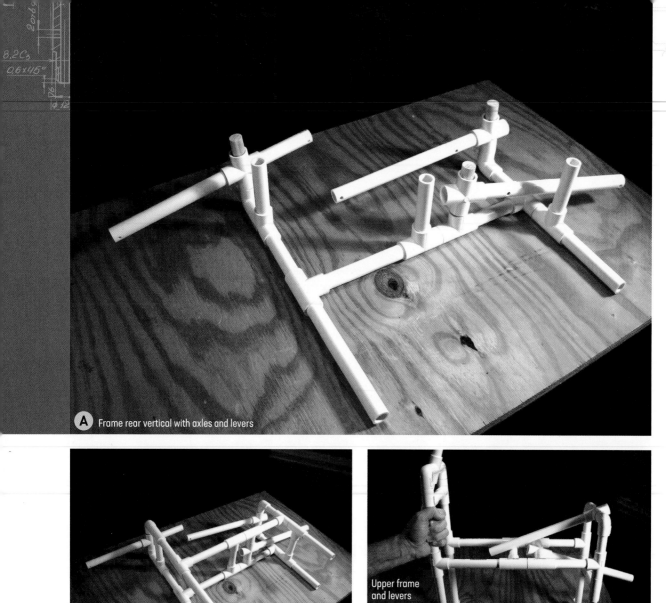

A Frame rear vertical with axles and levers

B Front and rear frames with axles and levers

Upper frame and levers joining with base frame

C

Upper frame and levers joining with base frame

D

Complete frame with levers and chains

E

Now we have completed the sub-assemblies that make up the frame.

The overall assembly of the Classy Lever Contraption is relatively straightforward and quick. Start by laying the rear frame assembly on its back as shown in Figure Ⓐ.

Place each of the three levers into the rear frame, pinning them in place with their dowel rod axles. Make sure that you install the levers in their correct locations and the correct orientation (Figure A). With the rear frame still lying flat on its back, install the front frame section by lining up the three axles and three spanning cross members. Once aligned, press and tap the front and rear frames together at the three cross members (Figure Ⓑ). Now, assemble the section you just built into the base section, with the base section flat and in its upright position (Figures Ⓒ and Ⓓ). Make sure that you properly seat the vertical members in the vertical sockets of the base frame.

The only thing left to do is link the three levers together and provide for ways to attach weights to the input side and output side of the contraption. For this final part of the final assembly, refer to Figure Ⓔ. Each of the drilled holes in the individual lever pipes needs to have an eyebolt installed in it. Referring to Figure E, ensure that you install the eyebolts in the correct orientation. For now, how far they stick out from the pipes is not that important, since we adjust how much they stick out later. For more information on installing eyebolts, see *Staying on Track: Countersinks and Eyebolts* on page 126.

With all of the eyebolts installed, it is time to attach the chains. Using pliers, bend out the uppermost chain link and the lowermost chain link far enough to get the links over the eyebolts (just as you did when you were creating your two chain sections of four links each at the beginning of the project). Now, again referring to Figure E, hook the open ends of the chain over the eyebolt sticking down from the Class 1 lever and the eyebolt sticking up from the Class 2 lever. Using pliers, close the open links to secure the chains on the eyebolts. Repeat this to link the Class 2 and Class 3 levers together. With all three levers linked to each other via the two internal chains, we now have a complete system!

Now we need to make some simple adjustments to

Ⓕ Complete classy lever project with water bottle weights

complete the Classy Lever Contraption. While holding the Class 1 lever in a horizontal position, adjust the length of the eyebolts in either or both of the Class 1 and Class 2 levers such that the Class 2 lever is also aligned horizontally. While continuing to hold the Class 1 lever in a horizontal position, repeat the adjustment procedure between the Class 2 and Class 3 levers. At this point, if you hold the uppermost, Class 1 lever horizontal and exert just enough force on the end of the lowest, Class 3 lever to take all of the slack out of the two linking chains, all three levers should be horizontal. Congratulations! You have completed the Classy Lever Contraption! Now it is time to utilize it.

To utilize the Classy Lever Contraption, add some weight to the load point eyebolt of the Class 3 lever. We found that a couple of small water bottles worked well (Figure Ⓕ). If you want to calculate a mechanical advantage for your contraption, weigh your water bottles prior to hanging them. Note that the load point on the Class 3 lever is also the system load point for the overall Classy Lever Contraption. Now, add weight to the input end of the system. This is the effort point of the Class 1 lever shown on the left of Figure F. Your goal here is to balance the system by adding enough weight to the effort end of the system to lift the weight at the load end. The Classy Lever Contraption is in balance when all of the levers are horizontal or parallel to each other.

So, how much weight on the effort end of the system did it take to lift the weight on the load end and balance the system? How close did you get to the effort or input load you calculated for your contraption from the *Tracking Further: Classy Lever Contraption in Math Form* (on page 118)? To calculate the mechanical advantage of your Classy Lever Contraption, divide the weight you added to the load end of the system by the weight on the effort end of the system required to balance it. Was the mechanical advantage of your Classy Lever Contraption close to what you expected? ❷

Staying on Track: Eyebolt Adjustment Made Easy

A simple and effective way to tighten a small eyebolt is to use an ordinary nail through the eye to keep it from rotating, while using a nut-driver to tighten the nut.

Eyebolt adjustment trick

Staying on Track: Countersinks & Eyebolts

A countersunk hole (also known as a countersink) is a quick and simple way to recess an eyebolt. At times, you may need to allow the base of an eyebolt's eye to be adjusted a short distance within a pipe's outside diameter. This trick is also useful in recessing an eyebolt within a flat surface, such as the face of a 2×4.

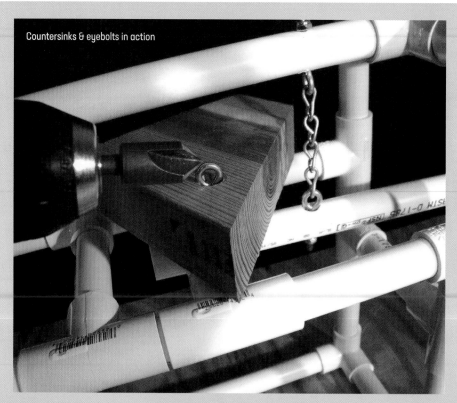

Countersinks & eyebolts in action

It is difficult to drill multiple, evenly-spaced holes in a straight line. Whether lines of rivets on an airplane's skin, or holes for mounting a tank's suspension arms, alignment of your holes can make all the difference in how well your machine works *and* looks. Marking correct hole placements is especially hard to do if you have a surface that is larger than any ruler you may have. Something learned from working on aircraft is the use of the "accordion tool," which takes the idea behind a scissor lift and extends it a little. This tool is available pre-made from several suppliers, but this project shows you how to create your own.

HOW IT WORKS

The accordion is nothing more than a series of interconnected levers, with holes drilled at the same spots on each arm of the scissors. The central holes are where two scissor arms meet to make an X, and the holes at the far edges are where the X's are joined together. Each arm in the X is the same, except one arm has an extra "marking" hole between the center and edge holes. When all these X's are joined together, every other arm will have a hole midway between the center and the top holes that are the same distance apart. If you set the hole spacing between any two holes, that spacing will be consistent, and in a straight line for all of the marking holes. The number of holes you can mark is the same as the number of X's you have joined together (Figure A). You can mark additional holes by simply aligning two already-marked holes on any two marking holes on the accordion, and continue marking until you are finished.

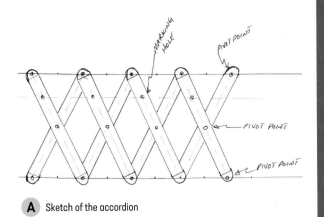

A Sketch of the accordion

CONSTRUCTION

The size of your X determines how much spacing you are planning to address with the accordion. The bigger the X, the wider the spacing you can mark. We will make a small accordion based on wooden craft sticks (Figure **B**), which are available in your local craft store and are just shy of 6" (148 mm) long.

Choose an even number of sticks (we chose 12). The number of holes you can mark is one half that number, or in this instance, 6. We suggest fabricating more sticks than you think that you will need, just in case you have any casualties when putting this together.

B Wooden craft sticks

Materials:

- » **Wooden craft sticks, 6" long (or 148mm x 20mm), 12–20 pieces**
- » **M3 × 0.5mm × 10mm screws or #6-32 × 3/8" screws**
- » **M3 (0.5 mm pitch) Nyloc Nuts or #6-32 Nyloc Nuts**

Tools:

- » **Drill Press or Hand Drill**
- » **⅛" or 3mm Drill Bit**
- » **Screwdriver or Allen Key** (to fit your screw type)
- » **Clamp**
- » **Pen or Fine-Tipped Marker**
- » **Ruler with Metric Measurement Scale**

STEPS:

1. Separate your pile of sticks in half. We chose 12 sticks, so we had two sets of six.

2. Select one stick from the first set and mark a hole dead center. This means this hole will be 74mm from the top and 10mm from the side.

3. Mark two holes 15mm from the top and bottom tip of each stick.

4. Mark one more hole about mid-way between the center hole and one of the tip holes. This becomes the location of the "marking" hole.

Staying on Track: Hole Position

The actual position of the marking hole governs the range of spacing your tool can handle. We chose 30mm from the stick's tip hole. With this location of the marking holes, our tool works over a hole spacing range from approximately 50 mm (~2") to 112 mm (~4 ⅜")... not bad!

5. Drill all four holes. This is now your marking stick to use as your template for marking the remaining sticks.

6. Stack the first set of sticks with the template on top, and clamp them together. Secure the stack on the drill press, or if you don't have a drill press, secure them to a surface that you can drill into.

7. Being careful to drill straight down, use the template to drill the four holes through the entire stack at once (Figure Ⓐ).

8. Repeat the process, but this time put your marking stick template on top of the second set of sticks and drill only the center and tip holes.

9. Now assemble the sticks by alternating sticks from each set. Put down all of the sticks from the second set diagonally, as shown, and then lay the sticks from the first set on top of the second set diagonally the other direction, forming X's all the way across (Figure Ⓑ).

10. Once all sticks are placed and forming X's, use screws and nuts to connect the sticks in the center holes and intercepting tips, connecting the sticks, as shown in Figure Ⓒ.

Staying on Track: Accuracy

Despite your best attempts, you are going to find that some holes do not appear centered no matter how hard you try, even though your marking stick looks perfect. This is because the fabrication process of the wooden craft sticks allows for slight dimensional variations. Welcome to the world of "tolerance." These sticks have loose tolerances, but this is not a limiting factor, because all our sticks' holes will be in perfect alignment and spaced the same relative to each other. The loose tolerances, in this case, do not impact the result. This is an important concept because some raw materials have large variations that you need to take that into account in your projects.

Now, as you stretch the accordion out along a straight line, you can mark your holes and each will be spaced the same distance apart, and in a straight line (Figure Ⓓ). ✷

A Stack of sticks on the drill press

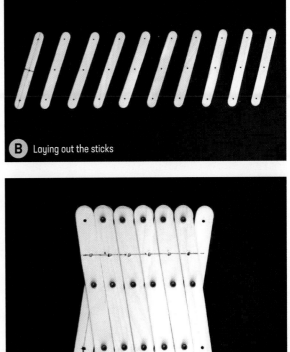

B Laying out the sticks

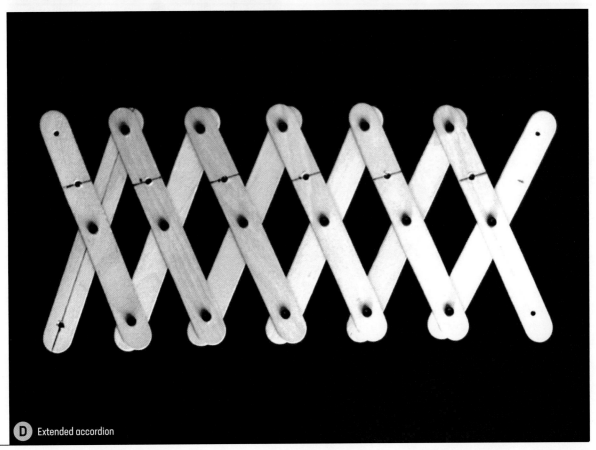

C Assembled accordion

D Extended accordion

Tracking Further:
Adult-Size Tank Suspension Component – Trailing Link _____

A Aircraft trailing link landing gear

You don't need to have a suspension system if you do not expect to go over any large bumps and don't mind a relatively rough ride. Tracked vehicles, because so much of their drive elements (the tracks) make contact with the ground, are susceptible to large oscillations (like "porpoising") whenever they encounter an obstruction. This causes whatever part of the track that makes contact with the bump to bounce up, taking the tank with it. Do this at speed, and you will have quite an uncomfortable ride, not to mention the likelihood of flipping over if you go fast enough against a big-enough bump. Worse, because no bump is nice and curved, any vertical surface your track encounters is going to feel like you slammed into a wall before you get pitched upwards.

This concept was understood long ago, when leather and then spring suspension was added to horse-drawn carriages. A very common use of the "trailing-link suspension" system today is on aircraft landing gear (Figure **A**), but it was also used in some early tanks from World War I and World War II (Figure **B**).

Samer attempted to prototype the same idea for the next version of his tank using some flanged bearings, a child's bicycle shock absorber found on eBay, and some 3D printed parts.

The trailing link suspension is really a system of levers that includes a mechanism to absorb abrupt movements, typically using some sort of strut or shock absorber (just as with many vehicles).

If you look at tank suspension image in this section, it's easy to see where Samer got the idea (Figures **C** and **D**). This idea evolved, and continues to evolve, as Samer devises the next version of his Adult-Size Tank.

B Tank suspension

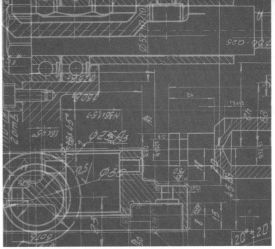

Levers are among the most basic, but most prolific and useful simple machines on earth. There are three classes of levers that provide differing functions. Levers can have a mechanical advantage equal to 1 (i.e., equilibrium), greater than 1 (amplifying force) or less than one (amplifying motion). They are used in so many ways and are pretty much everywhere. We hope that this discussion has heightened your appreciation for this simple machine, and we are sure that you will find many applications for levers in your maker projects.

C Aircraft trailing link landing gear

D Second version of Samer's trailing link suspension

7

Pulleys – Get into the Groove

In this chapter we discuss a wonderfully useful simple machine: the pulley. Archimedes (who we already mentioned in the levers chapter) is credited with developing the first fully-realized and documented compound pulley system. Plutarch, a Greek historian, wrote that Archimedes moved a naval ship at the request of King Hieron of Syracuse (in Sicily) by setting up a system of pulleys and then, while seated "some distance away," moved the ship "as smoothly and evenly as if she had been in the sea."

In its typical form, a pulley is a grooved wheel on a supported axle, around which the wheel rotates. Figure **A** shows a basic single pulley. The grooved pulley wheel holds a rope or cable under tension.

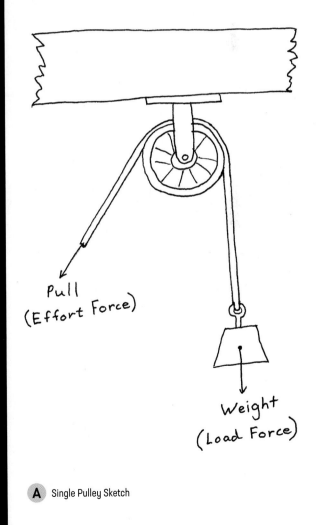

A Single Pulley Sketch

A Sketch of pulleys used in a water well

A single-pulley machine changes the direction of the applied effort force (or tension) relative to a given load or weight being lifted or moved. Take, for example, the task of lifting a pail of water out of a well (Figure **A**).

It would be incredibly difficult and awkward (not to mention dangerous) to try to position yourself directly over the well so that you could lift the bucket straight up. However, a pulley can come to the rescue. By mounting a single pulley to a structure directly over the well, a rope can be routed and pulled down or outward, at a convenient angle, to lift the bucket of water vertically out of the well. The single pulley provides a change of direction between the applied effort and the load, and therefore qualifies as a simple machine!

Recall from the lever chapter that mechanical advantage is calculated by dividing the load force by the effort force required to move the load. For a single pulley treated as an ideal, frictionless machine, the mechanical advantage is one. The system is only changing the direction, so the effort force is equal to the load force.

To obtain a mechanical advantage greater than or less than one (i.e., force amplification or reduction, respectively), we must use a compound pulley system. A compound pulley system is a machine comprised of a grouping of pulleys. The simplest compound pulley machine is made of just two pulleys working together and is known as a gun tackle (Figure **B**). Assuming that this gun tackle is an ideal machine, one with no friction, the mechanical advantage equals two.

To calculate the effort required in a compound-pulley system, divide the weight to be lifted by the mechanical advantage. For example, if a 20lb weight were to be lifted using an ideal gun tackle, the required lifting force or tension in the rope would be 10lbs (or half of the load force).

As mentioned with levers, there is always a tradeoff for mechanical advantage. A lever with a mechanical advantage of two requires an effort force half that of the load force. However, the effort end of the lever has to be moved two times further than the load end. The same is true for pulleys, and is called displacement. To calculate

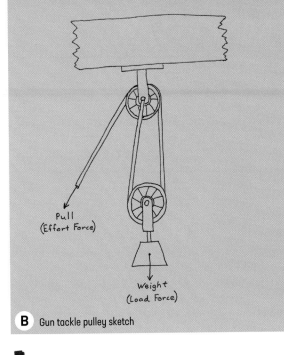

B Gun tackle pulley sketch

Labels in figure: Pull (Effort Force); Weight (Load Force)

C Tensioning pulley on Tot-Size Tank (side view, track removed)

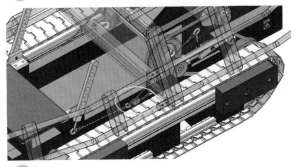

D Tensioning pulley on Tot-Size Tank (oblique view, track intact)

E Tensioning pulley on Tot-Size Tank (photo)

the displacement of the rope being pulled in the system, multiply how far the load is lifted by the mechanical advantage. So, in our ideal gun tackle example mentioned above, for every foot the load or weight is lifted, the rope being pulled must move two feet.

In addition to changing direction, pulley systems also amplify force. Consider the rope that is used to transmit the tension force around the pulleys. The rope is a single, continuous entity that ultimately supports the weight being lifted or moved. In the ideal gun tackle example (Figure B), the weight is attached to a movable pulley, which is supported by the rope. The rope is under a single tensile force. Two sections of the rope support the movable pulley, so the tension in the rope is only half the weight being lifted.

If the pulley system is in static equilibrium (that is, not accelerating or decelerating), the forces must sum to zero, in other words, they cancel each other out. For a given ideal pulley system, the tension required to lift a given weight can be calculated by dividing the weight by the number of rope segments supporting a movable pulley or pulleys.

The Tot-Size Tank uses a pulley as part of the control system for tensioning each track drive belt (see Figures **C**, **D**, and **E**). In this particular case, the pulley amplifies the motion of the tensioning arm relative to that of the control handle. The force generated by this system is effectively cut in half as the output displacement is double. This pulley system has an MA less than one (0.05).

These concepts of mechanical advantage versus the effort required to lift a weight, and the necessary displacement of our lifting rope, will become much clearer as we experiment with some simple pulley machines.

Claw and Pulley Experiments

You will find that many times in making, you must pause and detour your current project to come up with a device to help you experiment with the subject matter in which you are interested. For pulleys, we need to utilize a support frame to hold the pulley systems during testing, since pulleys lack their own support. The door frame molding in your home is an excellent basis for a support frame for our pulley experiments. Now, we need to make a contraption to simply (and temporarily) hold pulleys in a door frame. Hence the **Door Frame Clampy Claw** (DFCC).

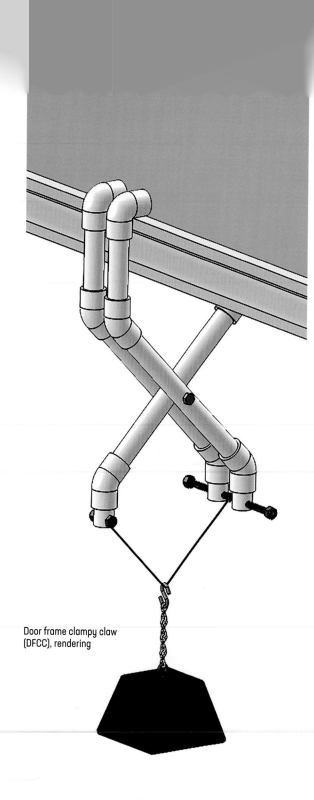

Door frame clampy claw
(DFCC), rendering

This nifty contraption is a self-locking lever mechanism that has a similar look and function of a "claw" made of a thumb and two fingers. Three identical "finger" structures make up the claw. Each finger is made of three lengths of PVC pipe, two 45° PVC elbow fittings, and one 90° PVC elbow fitting. The fingers are secured together with a single bolt acting as the fulcrum, allowing the fingers to open and close. When clamped on the wall, the "fingertips" touch the door frame in three places: two on one side of the door frame, and one on the other. The placement of the fingertips on the frame keeps the forces on a single plane and prevents twisting. The finger parts below the fulcrum provide an area to attach a cord that in turn will be used to hang from the claw whatever you plan to support. From this cord, we will hang the pulleys in our pulley experiments to come.

Note that the DFCC is made up of Class 1 levers, as presented in Chapter 6. This is an example of a practical way to use levers to make a useful, yet simple mechanism to perform a specific task.

CONSTRUCTION

To start the build of the DFCC, refer to Figure Ⓐ, which shows an exploded view of a single finger. Cut the PVC pipe to the lengths specified and then drill the holes as shown in the image. You will be making three identical fingers, so you will need to cut and drill enough pipe for all three. Assemble each finger as shown in Figure Ⓑ, paying careful attention to the orientation of the holes.

At this point, you are ready to connect the three fingers via a common fulcrum or axle. The axle is simply a ¼"-20 × 3" long bolt. Take one of the finger assemblies and insert the bolt through the center hole as shown in Figure Ⓒ. Take a second finger assembly (flipped over relative to the first one) and slide over the axle bolt in the orientation shown (Figure Ⓓ). Now slide the last finger over the axle bolt oriented in the same orientation as the first finger assembly (Figure Ⓔ). Thread a ¼"-20 nyloc nut onto the axle bolt and tighten until the assembly is snug, but not so snug as to crush the pipe. The finger assemblies should be able to be rotated relative to each other with some noticeable friction.

Materials:

- » **PVC Pipe – ½" x 5' – standard schedule 40**
  (see cut list for details), **1 length**
  (MMC* part # 48925K91)
  - **Cut to 9⁵⁄₁₆" length, 3 pcs**
  - **Cut to 4" length, 3 pcs**
  - **Cut to 1½" length, 3 pcs**
- » **PVC Fitting – Elbow (90°) – ½", qty 3**
  (MMC part # 4880K21)
- » **PVC Fitting – Elbow (45°) – ½", qty 6**
  (MMC part # 4880K31)
- » **Waxed Cotton Cord, 1.5 mm (~¹⁄₁₆"), 25yds, 12"**
  (Hobby Lobby part # 186494)
- » **¼"-20 Threaded Rod, 6", 1 length**
  (MMC part # 98750A436)
- » **¼"-20 × 3" Hex-head Bolt, qty 1**
  (MMC part # 91309A554)
- » **¼"-20 × 1¼" Hex-head Bolt, qty 1**
  (MMC part # 91309A544)
- » **¼"-20 Nyloc Nuts, qty 5**
  (MMC part # 90640A129)
- » **S-hooks, qty 2**
  (MMC part # 9381T21)

Tools:

- » **Saw** (either handsaw, miter saw, or PVC pipe cutting tool)
- » **Pliers**
- » **Drill**
- » **¼" Drill Bit**
- » **Rubber Mallet**
- » **Fine-Tipped Marker** (such as a Sharpie)
- » **Duct Tape**

* MMC = McMaster-Carr www.mcmaster.com

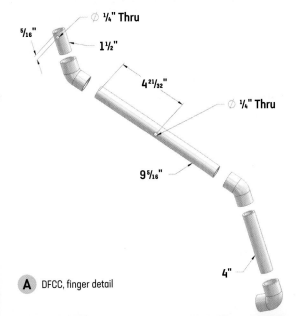

A   DFCC, finger detail

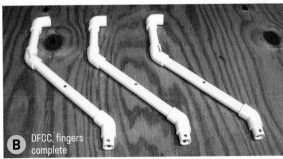

B   DFCC, fingers complete

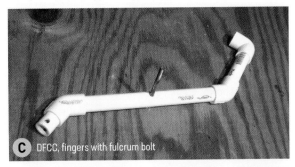

C   DFCC, fingers with fulcrum bolt

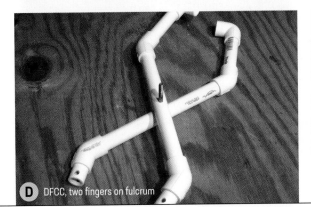

D   DFCC, two fingers on fulcrum

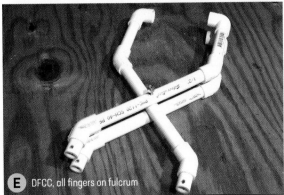

E   DFCC, all fingers on fulcrum

A DFCC, hanging end of the single finger

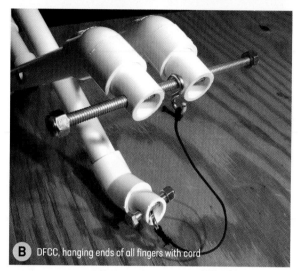

B DFCC, hanging ends of all fingers with cord

Staying on Track: Cutting Threaded Rod

When cutting a threaded rod, it is very easy to damage the threads in the area of the cut. It is then very difficult, if not impossible, to start a nut on the cut end. However, there is a simple way to prevent the threads from becoming permanently damaged.

Before cutting the threaded rod, thread a nut onto the rod beyond where you plan to cut it. Then after making your cut, unscrew the nut off of the rod over the cut end. Unscrewing the nut cleans up the threads if they are damaged, making it easy to start a nut on this cut end of the rod. You are effectively using the nut like the die in a tap and die set to re-cut the threads. If a sharp burr remains, you can file that away, leaving a small chamfer on the end of the rod to make reinstalling a nut easier.

Use a bolt to "clean" cut threaded rod

Next, we fasten a 12" length of cord to the lower end of the single finger and in the middle of the two fingers on the other side. This cord tightens the claw and can be used to hang whatever you plan to support from the DFCC. Insert a ¼"-20 × 1¼" long bolt through the small length of pipe at the lower end of the single finger assembly, but only push it through one of the through-holes. Before pushing it through the second hole, capture an S-hook on the inside of the pipe as shown. Fasten the bolt in place with a nut, and then tie one end of the cord to the free end of the S-hook (Figure **A**).

Now we install a threaded rod through the ends of the other 2 fingers. Start by cutting a 6" length of ¼"-20 threaded rod (*See Staying on Track: Cutting Threaded Rod* above). The length of threaded rod provides a way to connect the two fingers on this side of the claw together, and it supports an S-hook holding the other end of the main hanging cord. Insert a 6" length of ¼"-20 threaded rod through the hole in the lower end of one of the fingers on the duel finger side of the claw. Feed an S-hook onto the threaded rod and insert it into the lower hole of the other finger. Center the threaded rod on the two fingers as shown in Figure **B**. Now that the threaded rod is in place, tie the other end of

the cord to the S-hook on the two-finger side such that the length of cord between the two S-hooks is approximately 7". To complete the DFCC, add nyloc nuts to each end of the threaded rod (Figure B). The fingers' geometry/holes should provide enough friction to keep the threaded rod in place without using nuts. But if you find that your threaded rod is too loose, you can add nyloc nuts to each side of the fingers. Wondering why the threaded rod is so much longer than the combined width of the two fingers? The length of threaded rod protruding from each side provides a convenient place to secure fixed cords or lines to the claw.

Figure **C** shows the fully-assembled Door Frame Clampy Claw. To mount the DFCC, widen the lower half of the mechanism, slip the "fingertips" above the door molding, and push it closed. Pull down on the cord between the threaded rod and bolt at the bottom of the claw to "set" it in place. As you can see in Figure **D**, the more force/ weight that is added, the tighter the claw holds to the doorframe. See *Tracking Further: Free-Body Diagrams and Force Balance Equations on page 140*, for a mechanical engineering analysis of how to calculate the forces in the DFCC mechanism. ◗

C Door frame clampy claw (DFCC), fully assembled

D Door frame clampy claw, installed with weight

Tracking Further:
Free-Body Diagrams and Force Balance Equations

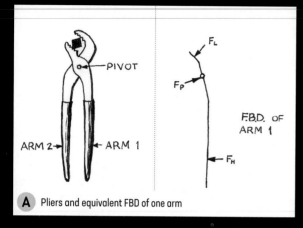

A   Pliers and equivalent FBD of one arm

Here we cover some of the things engineers do to understand how a structure or system is acted upon by forces. We go into some depth on a number of topics to give you a feel for how the process can work. But understand that there is far more to this than can be presented here.

When an engineer is analyzing the forces in a structure, mechanism, or other system, it is useful to have a simplified sketch of the system so they can be visualized. We call these simplified sketches **free-body diagrams**, or **FBDs**. The key elements of an FBD are:

1. A simple representation of the geometry of the system, like a stick figure.
2. Connections to other objects in the system.
3. Arrows showing the forces, both known and unknown.
4. Optionally, relevant dimensions and other information. But sometimes these are omitted to keep the diagram simple, as long as the information is listed nearby.

In Figure **A**, a detailed sketch of a common pair of pliers is shown on the left, with the corresponding FBD of one of its arms on the right.

In this FBD, the simplified shape of Arm 1 is drawn as a stick figure. The pivot connection to Arm 2 is represented by the circle. The force applied by the user is shown as F_H (known direction and magnitude). The force applied to the object in the pliers is shown as F_L (known direction, unknown magnitude) And finally, the load transmitted from Arm 2 to Arm 1 through the pivot is

shown as F_P (unknown direction and magnitude). (For this example, the dimensions are left off, but we can assume these are known.) This FBD allows us to quickly see the relationships between the various forces acting on Arm 1. From this, we can use what we call force balance equations to solve for the unknowns.

To talk about FBD analysis, a couple of new terms are needed. The term **vector** is used to describe something that has both magnitude and direction, like a force. As you'll see, it is not enough to know how much force is being applied. To fully understand its effect, we must also know the direction. It is possible, indeed likely, that you will know either the magnitude of a force, or the direction, but not both. Until both are known, the vector is not fully defined. By contrast, the term **scalar** is used to describe things that are only measured by a magnitude. For example, the area of a surface, the volume of a tank, or temperature of an object are all scalars.

Let's look at a different example and actually solve for some unknowns! Looking at the Door Frame Clampy Claw (DFCC) project in this chapter, it would be very useful to know the forces that it generates where it grips the wall. It would also be useful to know whether or not the central pivot is strong enough to withstand the load it undergoes. Using an FBD and force balance equations, that's exactly what we can figure out.

The DFCC is made of three identical fingers. The arrangement of the DFCC puts a single finger on one side of the doorway, and the other two paired together on the other side. Let's suppose we want to know all of the forces acting on the fingers. Where do we start? Let's start at the point where the known load is applied to the system, which is the weight hanging below the claw. Recall that it is hanging from a string suspended across the opposite fingers of the claw. Figure **B** is an FBD showing the forces applied at the point where the weight is hanging from the string.

In this FBD, we assume a coordinate system where the x-axis is to the right, and the y-axis is up. F_L represents the vertical load exerted by the item being hung from the DFCC. The magnitude and direction are known. F_A and F_B are the tension forces being applied by the string.

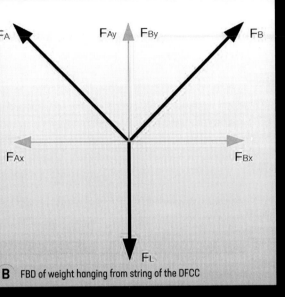

B   FBD of weight hanging from string of the DFCC

x- and y- axes, and a third for rotation. This particular FBD is one of the simplest possible, because all of the forces act upon a single point. So we can disregard rotation, leaving just two directions. Let's look at our force balance equations for this FBD:

$$\sum Fx = F_{Bx} - F_{Ax} = 0 \quad \text{Balance of forces along the x-axis}$$
$$\sum Fy = F_{Ay} + F_{By} - F_L = 0 \quad \text{Balance of forces along the y-axis}$$

Now, with a little algebra, we can figure out our forces. To begin, we should consider the heaviest weight we'd like the DFCC to carry. Given the weight of the toy used in the project, plus the pulleys and other items, let's use 10 pounds. This is F_L. Let's further assume that the string is sloped at 45°, and that everything is symmetric. We can use the force balance equations to find F_A and F_B:

From x direction balance: $F_{Bx} - F_{Ax} = 0$
$$F_{Bx} = F_{Ax}$$

From y direction balance: $F_{Ay} + F_{By} - F_L = 0$
$$F_{Ay} + F_{By} = F_L$$

From symmetry: $F_{By} = F_{Ay}$
$$F_{Ay} + F_{Ay} = 2 * F_{Ay} = F_L$$
$$F_{Ay} = \tfrac{1}{2} F_L = {}^{10}\!/_2 = 5$$
$$F_{By} = F_{Ay} = 5$$

From known geometry: $F_{Ay} = F_A * sin\ (\theta)$
$$F_A = \frac{F_{Ay}}{sin\ (\theta)} = \frac{5}{sin\ (45°)} = \frac{5}{0.707} = 7.07$$
$$F_{Ax} = F_A * cos\ (\theta)$$
$$= 7.07 * cos(45°)$$
$$= 7.07 * .707 = 5$$
$$F_{Bx} = F_{Ax} = 5$$

We know their directions, but not their magnitudes.

Before we go further, we need to do a little more work to prepare for calculation. Any forces (vectors, in general) not parallel to either the x- or y-axes need to be represented by their x and y components (scalars, in general). A component is the amount of a force parallel to an axis. If you look again at the FBD above, you'll notice that F_A is angled up and to the left. Alongside it are shown its components along the x- or y-axes. So for example, the x component of F_A is F_{Ax}, and the y component is F_{Ay}. We have to use some basic trigonometry to calculate a force's components. If we define the angle θ as the angle between the horizontal x axis and the direction of the force, the equations for the force components look like this:

$$F_{Ax} = F_A * cos\ (\theta)$$
$$F_{Ay} = F_A * sin\ (\theta)$$

This is a common way to reduce a vector to two scalars for subsequent analysis. There are other ways to do this, as well as analysis methods that work directly with vectors. But these are beyond the scope of this book.

To analyze this FBD, we use **force balance equations**. Force balance equations add up all of the forces in a certain direction and, for a non-moving or static system, sum them to zero. For 2D problems such as this, there are three force balance equations. There is one each for the

Tracking Further:
Free-Body Diagrams and Force Balance Equations Continued _____

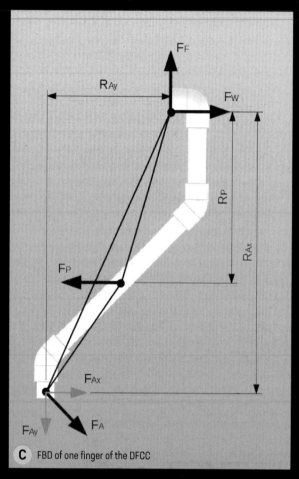

C FBD of one finger of the DFCC

Now, let's use this information to examine one of the fingers of the DFCC. Figure **C** shows an FBD of the single finger superimposed on a CAD model, with arrows representing the various forces applied to it, and dimensions relating their locations.

We'll use the same coordinate system in this FBD, just to help us keep everything oriented. The **x**-axis is horizonta pointing to the right. The **y**-axis is vertical, pointing up. Now, let's start with what we know:

● To keep things simple, we make some assumptions:
  • The mass of the finger is small enough to be ignored
  • Any friction in the pivot is negligible and can be ignored.

● We know these dimensions:
  • R_p is 7.63"
  • R_{Ax} is 12.61"
  • R_{Ay} is 5.69"

● We know these forces from our earlier work:
  • F_{Ax} is 5 pounds.
  • F_{Ay} is 5 pounds.

● We know there are three more forces, but they are not fully understood yet:
  • F_p is the force exerted between the sides of the claw through the central pivot. We know it is horizontal because of symmetry. We do not know the magnitude, or whether it points right or left. We will assume it is to the left. (You can assume either direction as long as you get the sign right in your forc balance equations!) If the magnitude turns out to be negative, we will know it is opposite our assumption.
  • F_w is the horizontal force exerted between the upper end of the finger and the side of the wall. We don't know its magnitude.
  • F_F is the vertical force exerted between the upper end of the finger and the door frame. We don't know its magnitude. Can you guess what it might be, knowing the whole DFCC supports 10 pounds?

We have three forces that are unknown in magnitude: F_p, F_w, and F_F. Just like before, we'll use force balance equations to calculate their magnitudes. Let's start with **x** and **y**:

$\sum Fx = F_W - F_P + F_{Ax} = 0$ Balance of forces along the **x**-axis
$\sum Fy = F_F - F_{Ay} = 0$ Balance of forces along the **y**-axis

Remember that if a force points left or down, it is negative. Lastly, we need to examine rotation, because unlike before, the forces are applied at different locations. The unique thing about rotation is that you have to specify the center about which the rotational forces (torques) are calculated. You could choose any point you want. But in this case, we want to pick a point where the fewest forces are known. Why? Because the torque applied by a force depends on the distance between where the force is applied and the center of rotation. By choosing the center at a point where forces aren't known, they can be ignored, because their torques are zero! With that in mind, let's choose the upper end of the finger. This lets us ignore F_W and F_F. Just as with **x** and **y**, we need to be mindful of direction. In this case, a force that tries to rotate our system counterclockwise is positive, while the opposite direction is negative. Here's the rotation balance equation:

$$\sum M = \left(F_{Ax} * R_{Ax}\right) + \left(F_{Ay} * R_{Ay}\right) - \left(F_P * R_P\right) = 0$$
Balance of torques

Now we have to use a bit of algebra to find F_P:

$$\left(F_{Ax} * R_{Ax}\right) + \left(F_{Ay} * R_{Ay}\right) - \left(F_P * R_P\right) = 0$$

$$F_P * R_P = \left(F_{Ax} * R_{Ax}\right) + \left(F_{Ay} * R_{Ay}\right)$$

$$F_P = \frac{\left(F_{Ax} * R_{Ax}\right) + \left(F_{Ay} * R_{Ay}\right)}{R_P} = \frac{(5*12.61) + (5*5.69)}{7.63} = 12$$

Note that since F_P is positive, we guessed correctly on its direction. Finding F_F is easy enough:

$$F_F - F_{Ay} = 0$$
$$F_F = F_{Ay} = 5$$

All that's left is F_W:

$$F_W - F_P + F_{Ax} = 0$$
$$F_W = F_P - F_{Ax} = 12 - 5 = 7$$

So what does this mean for the real DFCC? If you think about it, F_F should be exactly half of the force pulling down on the claw due to the weight hanging from it. (Did you guess that correctly?) And indeed, it is exactly half of our initial force of 10 pounds. At the same time, the central pivot that joins the fingers must support F_P, or 12lbs of force trying to pull it out of each finger. And if each finger is exerting F_W against the wall, the entire claw is compressing the wall with a total force of 2*7 = 14lbs. The claw exerts about 40% more force against the wall than the weight of whatever is hanging from it! One last thing to remember is that the other side of the Clampy Claw uses two fingers. So each finger will see half the force of the single one we just analyzed.

This example shows how an FBD and force balance equations can be used to calculate the forces in your structures and mechanisms. This can help you design joints, joints, and choose fasteners, bearings, and other items, to be strong enough for your project to function without premature failure. Remember, a good strategy here is to start with the forces you know, and work logically to eliminate the unknowns one by one. Using a good choice of rotational center allows you to disregard the most unknowns and solve for the rest. Also try to incorporate a check of your results, as we did with F_F, to make sure the results make sense. In the pulley examples later in this chapter, we make frequent use of FBDs to help visualize the forces acting on the system. It is always a good idea to sketch out an FBD any time you're doing this type of analysis. It does not need to be formal, like the ones I have drawn. As you'll see, a simple sketch will often do the trick. —*John Manly*

Materials:

» **Door Frame with Molding**, qty 1
» **Door Frame Clampy Claw (DFCC)**, qty 1
» **Paracord Rope, 15' (minimum)**
(MMC* part # 3696T38)
» **Fish/Spring Scale or Kitchen Scale** to measure up to 5 lbs, qty 1
» **Item weighing ~4 lbs** (ex. jug of water), **qty 1**
» **Item weighing ~2 lbs** (ex. plastic toy tractor), **qty 1**
» **Zip Ties**, qty 10+
» **Clothesline Pulley** (2" or larger is recommended), **qty 4** (Lowe's Blue Hawk part # 349199)
» **Double Pulley, qty 2** (MMC part # 3742T43)
» **S-hook** (optional, can use zip ties), **qty 2+** (MMC part # 9381T21)
» **PVC Pipe – 4" long × ½" diameter, qty 1** (MMC part # 48925K91)
» **Eye Bolt – #8-32 with 1⅛" shank length with nuts, qty 6** (MMC part # 9484T46)

* MMC = McMaster-Carr www.mcmaster.com

PARACORD is a very strong, lightweight nylon cord used in many applications where both strength and abrasion resistance required. Paracord is used for parachute lines; hence the origin of its name. Paracord is constructed as a kernmantle rope. Kernmantle rope has a central, strong core protected by an outer woven sheath.

Close-up of a kernmantle rope end

Now let's work with pulley systems to explore the mechanics of these simple machines firsthand. The following experiments will give you a more in-depth understanding of the power and uses of pulleys. Each experiment includes brief setup instructions, and illustrations to guide you through them. Each pulley configuration is supported in a door frame by the DFCC. Please be careful not to overload the claw; make sure to keep the combined weight hanging from the DFCC to 10lbs or less.

USING THE PULLEY EXPERIMENTS TO MEASURE PULL (EFFORT) FORCE

In each of the five pulley experiments described below, the goal is to measure the force applied to the pulley cord (effort force) to lift a known amount of weight (load force). The most direct and perhaps the best way to measure this pull force is to use a fish scale, also known as a spring scale. However, if you don't have a fish scale, a a small kitchen scale that can weigh up to 5lbs also works. Fill a milk jug with water to weigh 4lbs to use as one of your weights (if you are using a spring scale, use it in place of this water jug weight). The other weight (i.e., the weight to be lifted, or load source) should weigh about 2lbs (we use a toy tractor in the examples).

Once the experiments are set up according to the instructions, you can determine the amount of effort force required for that pulley system. To determine this, tie one end of the paracord to the tractor, and the other end to the milk jug. Place the milk jug on top of the scale, and measure the weight of the water jug WHILE the tractor is being lifted by the pulley system. Subtract the new (lesser) water jug weight from the original 4lb water jug weight. This calculated value is the pull force (Figures **A** and **B**).

Here is an example: Let's say that the scale reads 2lbs when the tractor and the milk jug are both connected to the pulley system. The milk jug weighs 4lbs. So if the scale reads 3lbs, then the pull force is the difference, or 1lb. This process becomes clearer once you begin the following experiments.

Measuring pull force
of a pulley system

A

B

PULLEY EXPERIMENT 1 (MA = 1)

Our first pulley experiment illustrates only a change in
direction without any reduction or increase in pull force. In
other words, the pull force should be equal to the weight
of the load, giving us a mechanical advantage of 1, as
illustrated in Figure **C**.

This pulley system is the simplest to make. First (as with
all of the pulley experiments), start the pulley system
configuration by mounting the DFCC in the door frame.
Then, using a zip-tie and an S-hook, hang a single
clothesline pulley from the lower cord of the DFCC (Figure
D).Once the pulley is secured to the DFCC, thread a length
of paracord over the pulley.

On the load end of the paracord, attach a hook by which
you can hang the toy tractor load. Tie the other end of the
cord to the milk jug weight, or a spring scale. If you do use
the milk jug weight, make sure that the length of the cord,
when it is tied to the milk jug, is such that the load that you
are lifting hangs in the air approximately halfway down the
door opening when the jug is placed on the scale.

Since the 2lb toy tractor is used as the load, the scale
should read close to 2lbs (4lbs of milk jug weight minus
2lbs of tractor weight). Note that we say "close to" 2lbs.
Considering that no mechanical system is ever perfectly
ideal (in this case perfectly frictionless), there is always
some small amount of energy lost due to friction. As
mentioned before, this system provides a mechanical
advantage of 1: the load force equals the pull force. The
advantage of using this system is that the motion of the
load changes direction.

C Experiment 1 - single pulley - change in direction (MA=1)

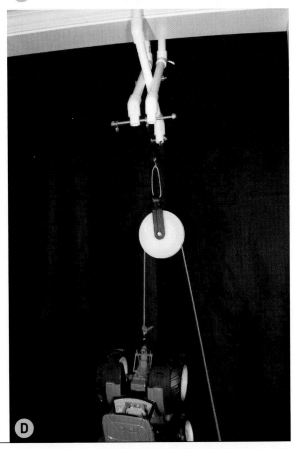

D

PULLEY EXPERIMENT 2 (MA = 2)

This experiment allows observation of both a change in direction and force amplification. You construct a pulley system that results in an ideal mechanical advantage of 2, as illustrated in Figure **A**.

As with experiment 1, we start by hanging a single clothesline pulley from the DFCC. Next, take your paracord and thread it through the pulley you've just hung. Then, take another clothesline pulley and thread the same cord through it as well. Next, take the free end of the cord and tie it to one side of the DFCC's threaded rod. Finally, connect the load to be lifted to the lower, free pulley. (See Figure **B** and **C** for the complete assembly.) Note that when you pull on the cord, it takes significantly less effort to raise the load. However, also note that for every 2" of cord you pull, the load is only raised 1". Assuming that your load is about 2lbs, you will see that the scale reads about 3lbs, which equates to a pull force of 4-3 = 1lb. This means that the pull force is roughly half that of the load being lifted; hence the fact that this configuration yields a mechanical advantage of 2.

The following three experiments show different pulley configurations that all result in a mechanical advantage of 4. In other words, each of the experiments below should produce a pull force approximately one-quarter that of the load being lifted. Also, note that it will take four times the length of rope being pulled in order to lift the load a given amount.

PULLEY EXPERIMENT 3 (MA = 4)

This experiment incorporates two single pulleys fixed to a rigid frame and two pulleys that are allowed to move. The two movable pulleys are constrained such that they move as a single, linked entity via a rigid bar. The load to be lifted is then attached to the rigid bar at its center. This is our first pulley system that should result in a mechanical advantage of approximately 4, as illustrated in Figure **D**.

To construct this experiment, we start by making the rigid bar that is used to support the two movable pulleys and the load to be lifted. Referring to Figure **E**, cut a piece of ½" PVC pipe to 4" long. Then drill the holes as shown and loosely mount the eye bolts using nyloc nuts. This gives us a handy means by which to connect the two pulleys and the load.

A Experiment 2 - two single pulleys (gun tackle) - change in direction/force amplification (MA=2)

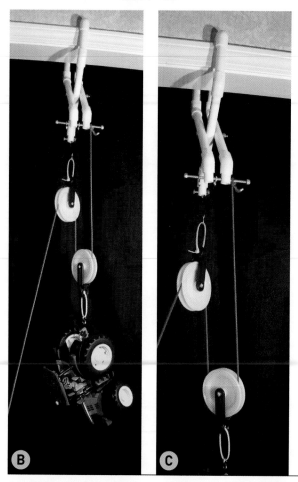

B **C**

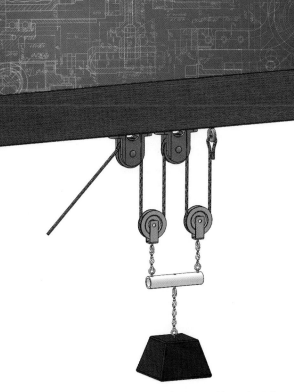

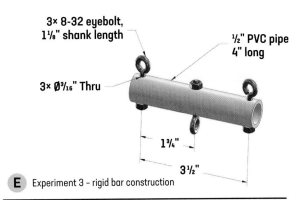

3× 8-32 eyebolt, 1⅛" shank length

½" PVC pipe 4" long

3× Ø⁹/₁₆" Thru

1¾"

3½"

E Experiment 3 - rigid bar construction

D Experiment 3 - two fixed and two moveable single pPulleys - change in direction/force amplification (MA=4)

As we have done previously, attach a single pulley to the lower cord of the DFCC. Next, hang another single pulley from one side of the DFCC's threaded rod. At this point, attach two more single pulleys to the rigid bar connecting them to the two upward-facing eye bolts. See Figure **F** for proper orientation.

The attachment of the pulleys to the eye bolts can be a chain as shown in Figure F, or you can simply use zip ties, as shown in Figures **G** and **H**. Although the method of attachment can vary, ensure that the pulleys are attached such that they are each an equal distance from the bar.

It is now time to thread some paracord through the pulleys, connecting them to make our pulley system. Start by threading the cord through one of the two pulleys attached to the claw. For consistency with our prior experiments, let's start with the pulley you hung from the DFCC's lower cord. Next, thread the cord around one of the movable pulleys attached to the bar. From here you need to take the cord over the second pulley attached to the claw. Now thread the paracord over the second movable pulley on the bar. Finally, tie the loose end of the cord to the claw's threaded rod on the side opposite where you attached one of the single pulleys. Whew! Our experiment 3 pulley system is finally complete.

Single pulleys

Movable pulley rigid bar

F Experiment 3 - rigid bar construction with attached pulleys

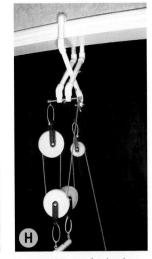

G Experiment 3 - two fixed and two moveable single pulleys - change in direction/force amplification (MA=4)

H Experiment 3 - two fixed and two moveable single pulleys - Pulley Orientation Detail

Now hang whatever load you are experimenting with (we are still using the 2lb tractor) and see what pull force is required (ours is approximately 0.5 pounds). Notice how far you need to pull the cord to raise the load a given amount.

PULLEY EXPERIMENT 4 (MA = 4)

In this experiment, you create a pulley system that uses three single pulleys and two separate lengths of cord to once again result in a mechanical advantage of 4. If you refer to Figure **A**, you can see that there are two movable single pulleys, connected by a different cord than the cord that is being pulled. The upper movable pulley is supported by a length of cord that is then routed through a fixed single pulley. This cord is the one pulled to actuate the system, and lift the load.

The first step in constructing this experiment is to attach a single pulley to the lower cord of the DFCC. Thread paracord over the pulley you just mounted, and then thread the same cord over another single pulley. Next, take the free end of the cord and tie it to one side of the claw's threaded rod. Your pulley configuration should now look like it did in experiment 2. Now we need to secure a second movable pulley below the first with another length of paracord. You may need to experiment a bit with the length of this second cord; however, a good starting point is 48". Start by tying one end of the second cord to the upper movable pulley. Now loop the loose end of the cord through another single pulley. This pulley becomes our lower, second movable pulley in the system. Tie the loose end of the cord to the claw's threaded rod on the same side that the other cord was tied. Finally, attach the load you plan to lift to the lower movable pulley. The final system is shown in Figures **B** and **C**.

While playing with this pulley system, note the pull force required to lift your load and the amount you have to pull the upper cord to move the load a given amount. The results should mimic those of the previous experiment. However, it is interesting to observe the movement of the second rope as compared to that of the first. Try measuring the movement of the second rope and the movement of the load as you pull the upper rope a known amount. What do you see?

A Experiment 4 - three single pulleys - change in direction/force amplification (MA=4)

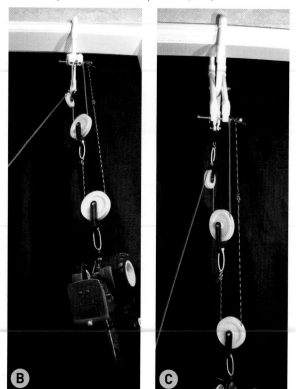

B Experiment 4 - three single pulleys - change in direction/force amplification (MA=4)

C Experiment 4 - three single pulleys - pulley orientation detail

PULLEY EXPERIMENT 5 (MA = 4)

In this final experiment, we introduce the double pulley. This is one of the simplest forms of a block and tackle. In essence, a double pulley is simply two single pulleys that share a common axle and frame (Figure 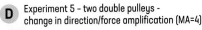). Every single pulley contained within a double pulley can spin together or independently, as well as in the same or opposite directions. In this experiment, observe how the use of two double pulleys makes for a nice, compact way to make a pulley system with a mechanical advantage of 4.

This is one of the simpler pulley systems to make. Start by attaching a double pulley to the lower cord of the DFCC. Next, secure another double pulley via paracord. Thread paracord through one side of the upper double pulley. Then thread the cord through one side of the lower double pulley. Now bring the cord back up to the upper double pulley and thread it through the unused side of the pulley. Take the cord back down to the lower pulley and thread it through the unused side of this pulley. Finally, take the end of the cord back up and tie it to one side of the DFCC's threaded rod. Now, hang the load to be lifted from the lower double pulley (Figure **E**).

The results here are very similar to the previous two experiments. However, you may note that this system does not come as close as the others to the ideal mechanical advantage of 4. This is because the pulley wheels within the double pulleys offer a good bit more resistance to rotation, because they are significantly smaller in diameter than the clothesline pulleys we use in all of the other experiments. This greater resistance in small pulleys is due to two main factors. First, a smaller pulley requires the cord to bend and unbend more than it has to on a larger-diameter pulley, causing more energy to be required. Second, a smaller pulley has a smaller lever arm (or moment) about which to overcome the axle friction than a larger pulley. In other words, for a given amount of axle friction, the smaller the ratio of the pulley diameter is to the pulley's axle diameter, the more the pulley will offer resistance to rotation. ◢

D Experiment 5 - two double pulleys - change in direction/force amplification (MA=4)

Tracking Further: Pulley Mechanical Advantage

At first glance, pulley mechanical advantage calculations may look daunting. However, in this Tracking Further, we will be showing you how to tackle these calculations in a simple, straightforward way.

Let's start by noting the assumptions involved in making in our calculations. First, we assume that the pulleys involved are ideal, or in other words, frictionless. Second, the weight of the rope and pulleys can be considered negligible and therefore, not included in the calculations. Third, we assume that the ropes act on the pulleys at 180° angles (the reason will become clearer once we dive into the calculations below). Last, we assume that the rope used in our pulley systems does not stretch (in reality the paracord we use in our experiments does stretch a bit. However, with the loads that our paracord sees during these experiments, this amount of stretch should be minimal and therefore can be ignored). Making these assumptions significantly simplifies our calculations without skewing our results too much from reality. In mechanical engineering practice, it is common and often even necessary to make certain strategic assumptions when analyzing mechanical systems mathematically.

Recall from our previous discussion that mechanical advantage is calculated by dividing the load force by the effort force. For the following calculations, we use W to represent the weight or load force and T to represent the tension or effort force, giving the equation for mechanical advantage (**MA = W/T**).

For each of the calculations, first we need to draw the free-body diagram representing each of the pulleys in the system to be analyzed. The free-body diagrams show the forces on a specific pulley from either the pulley's attachment to a fixed point, the interaction of the rope pulley, the weight being lifted, or a combination thereof. The variable R represents the reaction force due to the fixed connection of a given pulley. An equation for each pulley free-body diagram is written by simply summing the forces in the y-direction (vertically) and setting them equal to zero. Positive y components will be those that act in the up direction on the pulley. One of the most important principles that you need to understand when analyzing pulley systems is the idea that a single, continuous rope has the same tension in it at any point in the system. This too becomes much clearer once we get into the examples below.

Let's start by looking at the simplest pulley system scenario, as illustrated in Figure Ⓐ. There is only a change in the direction of the section rope being pulled versus the direction of the rope section/weight being lifted. With only a change in direction, the mechanical advantage should be 1. The input effort or tension in the rope (the force required to pull the rope) is equal to the load (weight) being lifted. Given that the one rope in this system sees the same tension force on both the side being pulled and the side of the pulley with the weight hanging from it, the effort force T required to lift the weight is equal to the weight. For completeness, the reaction R force or forces should also be calculated. From the force equilibrium equations, we can see that the reaction force is twice that of the weight being lifted. In some cases, this reaction force can be very important to calculate, in that it directly shows you what the structure supporting the pulley must be able to hold. Considering that the effort force, in this case, is equal to the load force, it follows that the value of mechanical advantage is equal to one.

Our second example is a bit more complicated, but the assumptions, principles, and approach are the same as the first example. Again there is only one, single, continuous rope in this pulley system with the same tension throughout. Now, however, there are two pulleys to consider. The free-body diagram for both pulleys is shown in Figure Ⓑ. So, where do we start in analyzing this system? Look at the two free-body diagrams. The one for pulley B involves the reaction force and the rope tension; both of which are unknown variables. Now, looking at the free-body diagram for pulley A, we see that the forces acting on it are the tension in the rope and the load force or weight being lifted. Great news! In this case, we know the weight, so we only have one unknown variable, the rope tension. So this is where we need to start. Summing the forces for pulley A and setting them equal to zero, we see that it is quite straightforward to solve for the tension force in the rope. Armed with the tension T in the rope, we can then substitute it into the summation of the forces set equal to zero for pulley B, thereby getting the reaction force R. The tension in the rope is the pull force required to lift the weight. In this particular case, we calculate that the tension in the rope is half that of the weight being lifted. Therefore this system results in a mechanical advantage of 2.

Now let's look at a pulley system with three pulleys and two separate ropes (Figure **C**). Each rope can be considered a single, continuous rope with the same tension throughout its length. However, as we soon see, the tension in the two ropes is not the same. As in the prior example, we start the analysis with the pulley that contains only one unknown variable, which is the pulley that has the weight attached to it. In this example, we start with the third pulley, pulley A. The rope that passes over this pulley is the second, continuous rope in the system with its tension force represented by the variable T_2. Writing the force equilibrium equation for pulley A and solving for T_2, we see that the value of T_2 is half that of the weight being lifted. So, at this point, we see a mechanical advantage of 2. However, we are not finished with the overall system yet. Next, we solve for the pull force or tension in the first rope by evaluating the forces exerted on pulley B. The tension in the first rope in the system, the rope that will be pulled, is half of the tension in the second rope. Yet again another mechanical advantage increase of two. So, the overall mechanical advantage of the system is four. The effort or pull force will be ¼th of the load force or weight being lifted.

The following example illustrates a pulley system that results in a mechanical advantage of 4 with a single, continuous rope. Two double pulleys are joined, as shown, by the one rope. The top pulley is fixed, and the lower pulley is allowed to move. The rope is threaded through the pulleys as shown (Figure **D**), and the end opposite that where the effort or pull force is applied is also fixed. Solving the force equilibrium equation for the moveable pulley results in a T or pull force of 1/4th of the weight being lifted, thus a mechanical advantage of 4.

In the final example, we explore a way to get a mechanical advantage of 8 out of a pulley system (Figure **E**). In reality, this system would not be very practical; using compound pulleys, you can achieve the same mechanical advantage in a much more practical, simpler system. However, the system in this example is much easier to follow from the standpoint of analysis. The approach is the same as all of the previous examples, so we don't go into detail for this example. You are now well equipped to following this one yourself!

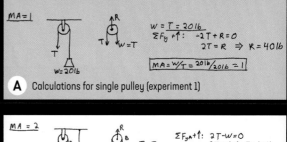

A Calculations for single pulley (experiment 1)

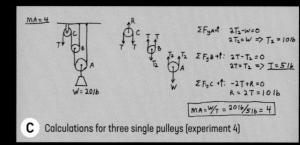

B Calculations for two single pulleys (experiment 2)

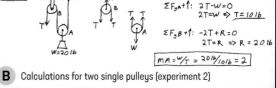

C Calculations for three single pulleys (experiment 4)

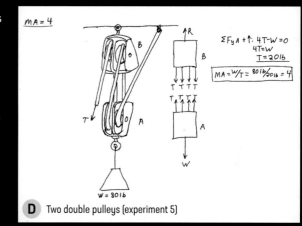

D Two double pulleys (experiment 5)

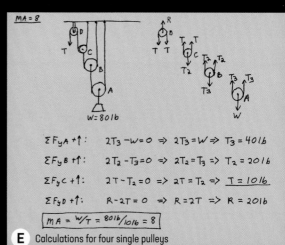

E Calculations for four single pulleys

The **PVC Articulated Crane (PAC)** project combines levers with pulleys to ultimately result in a fully functional, articulated toy crane. This build teaches you about utilizing pulleys in a more complex, real system by integrating them with two main levers and a base frame that make up the PAC.

WHAT MAKES UP THE PAC AND HOW IT WORKS

The PVC Articulated Crane is made up of a **base frame**, the main arm or **boom**, the upper arm or **jib**, and the **actuation pulley systems** (Figure Ⓐ).

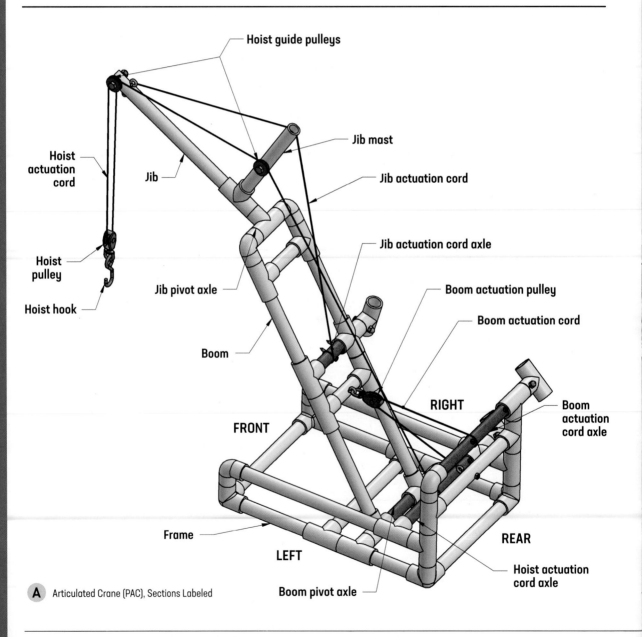

Ⓐ Articulated Crane (PAC), Sections Labeled

The base frame serves to support the pivot axle for the large boom and two of the three actuation cord axles.

1. The upper actuation cord axle is used to wind the cord connected to the boom actuation pulley system, thereby changing the angle of the boom.

2. The lower actuation cord axle raises and lowers the hoist pulley, cord, and hook system. This is directly connected to whatever object the crane will be lifting.

The boom's primary purpose is to support the jib and the jib actuation axle. The jib is connected to the boom via an axle that allows it to pivot about the top of the boom. The pivoting action of the jib allows for further articulation of the crane by adjusting its angle relative to the boom, but its ultimate purpose is to support the hoist guide pulleys and hoist cord. The control or actuation of the jib is achieved via a cord and axle system, and the jib actuation cord axle is part of the boom.

Let's get going with the build. The anatomy and workings of the PAC will become much clearer as we get into the build.

CONSTRUCTION

We build the PAC as four separate modules. The modules consist of the jib, the boom, the jib/boom assembly, and the base frame. Once we have built and assembled these modules, we add the cords, pulleys, and axles that actuate the boom and jib, and raise and lower the hoist hook.

We think that now is a good time for a bit of a disclaimer. The following instructions for building the PVC Articulated Crane are going to be far less detailed than what we've provided in the previous projects. However, as always, during the build of the PAC, we present several new tips and tricks. The instructions for this build are presented as a series of steps. Let's get started.

Materials:

» **PVC Pipe – ½" × 5' – standard schedule 40** (see cut list on the following page for details), **3 lengths** (MMC* part # 48925K91)

» **PVC Fitting – Elbow (90°) – ½" , qty 5** (MMC part # 4880K21)

» **PVC Fitting – Cross – ½", qty 2** (MMC part # 4880K241)

» **PVC Fitting – Tee – ½", qty 22** (MMC part # 4880K41)

» **PVC Fitting – Side Outlet Elbow – ½", qty 4** (MMC part # 4880K631)

» **¾" Oak (Hardwood) Dowel** (see cut list for details), **4' length, qty 1** (MMC part # 96825K19)

» **Waxed Cotton Cord, 1.5 mm (~¹⁄₁₆"), 16' length** (Hobby Lobby part # 186494)

» **¾" Single Swivel Eye Pulley (nickel-plated), qty 3** (Lowes part # Blue Hawk 656949)

» **Screen Door Slide Rollers - 1" diameter, qty 1 package with 2 pieces** (Lowes part # Prime-Line 55273)

» **#8 × ½" Wood Screws, qty 6** (MMC part # 90011A194)

» **#8-32 Eye Bolt, 1⅛" shank length, qty 3** (MMC part # 9484T46)

» **#8-32 × 1¼" Machine Screws, qty 6** (MMC part # 90272A201)

» **#8-32 Nyloc Nuts, qty 6** (MMC part # 90631A009)

» **¼"-20 × 1½" Hex-Head Bolts** (partially threaded), **qty 2** (MMC part # 91236A546)

» **¼"-20 Nyloc Nuts, qty 2** (MMC part # 90640A129)

» **Mulitiple Zip Ties**

» **Hook, qty 1** (see instructions)

Tools:

» **Saw** (either handsaw, miter saw, or PVC pipe cutting tool)

» **Pliers**

» **Drill**

» **³⁄₁₆" and ¼" Drill Bits**

» **Rubber Mallet**

» **Fine-Tipped Marker** (such as a Sharpie)

* MMC = McMaster-Carr www.mcmaster.com

Cut List

Qty	Material	Cut Length	Drill Holes (³⁄₁₆")	Module
2	PVC pipe	11½"		Frame
1	PVC pipe	10½"	Two through-holes	Jib
4	PVC pipe	9"		Frame
3	PVC pipe	6½"	Two through-holes (Jib only)	Jib (1), boom (2)
2	PVC pipe	6¼"		Frame
2	PVC pipe	4"		Boom
2	PVC pipe	4"	One through-hole (one only)	Boom
2	PVC pipe	3¼"		Frame
17	PVC pipe	1½"		Jib (1), boom (4), frame (1)
2	Hardwood dowel	12.5"	Two through-holes in each Two blind pilot holes in each	Actuation system
1	Hardwood dowel	7"	Two through-holes Two blind pilot holes	Actuation system
1	Hardwood dowel	4.5"		Boom
1	Cord	12"		Actuation system
1	Cord	24"		Actuation system
1	Cord	36"		Actuation system
1	Cord	120"		Actuation system

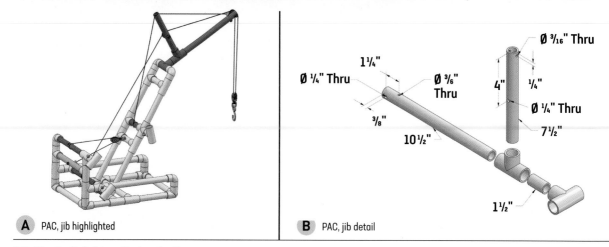

A PAC, jib highlighted

B PAC, jib detail

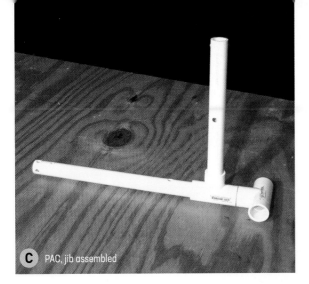

C PAC, jib assembled

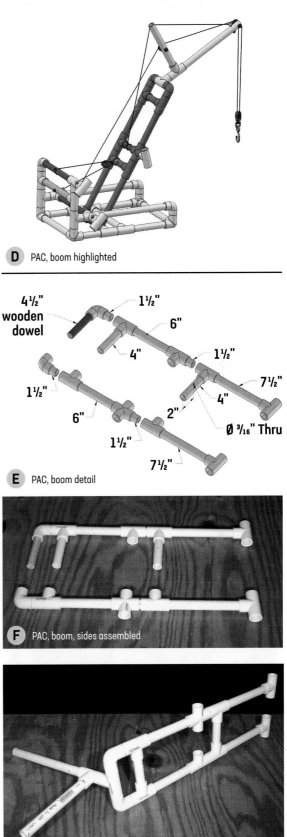

D PAC, boom highlighted

STEP 1

The first thing to construct is the jib (Figure **A**). Referring to Figure **B**, cut, drill and assemble the pieces to complete the jib module. The final jib assembly is shown in Figure **C**.

STEP 2

Now, construct the boom (Figure **D**). Referring to Figure **E** cut, drill and assemble as shown in the figure to build the boom module. You are building the left and right sub-assemblies of the boom. Be aware that the boom will not be completely assembled at the end of this step. The two final halves of the boom are shown in Figure **F**.

STEP 3

In this step, you mount the jib into the boom via a pivot dowel, completing the jib/boom module. Referring to Figure **G**, slide the jib over the pivot dowel already installed in the right half of the boom. Now, add the left boom sub-assembly to the right boom sub-assembly, trapping the jib pivot dowel (and thus the jib) at the upper end of the boom. Pay close attention to the orientation of the jib relative to the boom to ensure that the cross fitting for the jib actuation cord axle is on the correct side.

4½"
wooden
dowel

1½"

6"

4"

1½"

7½"

1½"

6"

2"

4"

Ø ³/₁₆" Thru

1½"

7½"

E PAC, boom detail

F PAC, boom, sides assembled

G PAC, jib and boom assembled

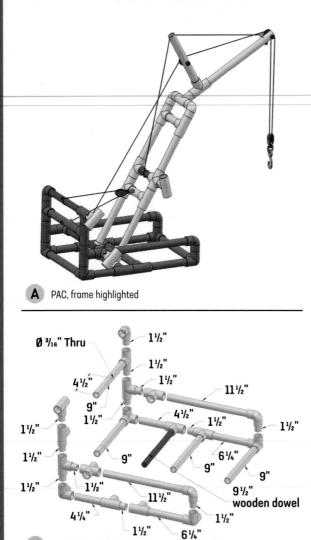

STEP 4

Now it is time to build the frame module (Figure **A**). The construction of the frame is done in two halves, just as was done for the boom in step 2. Referring to Figure **B**, cut, drill and assemble the parts for each half of the frame, as shown in the figure. The two final halves of the frame are shown in Figure **C**.

STEP 5

At this point, we are ready to assemble the jib/boom module into the frame module. Similar to step 3, slide the lower, pivot end of the boom onto the boom pivot dowel that should already be present in the right frame half (Figure **D**). Next, add the left frame half, trapping the boom pivot dowel and subsequently the jib/boom module (Figure **E**). The fundamental structure of the PVC Articulated Crane is now complete!

Now that we have built and assembled the primary structural components of the crane, it is time to make it move. The rest of the project entails hooking up various pulleys, cords and actuation cord axles.

STEP 6

Each of the three actuation systems (the boom, jib, and hoist actuation systems) require an axle by which you can reel in or pay out a cord. Referring to Figure **F** you are preparing all three axles with handles. Label each one (jib, boom, hoist) for later reference.

In an earlier design, we left the PVC Tee fitting off of the axle. The wooden dowel simply stuck out of the right side of the crane. It worked okay this way, but we found that the axles were much easier to turn if we added a PVC Tee fitting as a handle. To add the tee fitting handle, slide it onto one end of the wooden dowel as far as it will go. To make the tee fit more tightly to the dowel, you can wrap some duct tape around the dowel. This will have the effect of increasing the diameter of the dowel slightly so that you can achieve a nice, snug fit with the tee. Then, cross-drill a hole through the tee and the dowel. Secure the assembly together either using a wood screw or machine screw and nyloc or acorn nut. We used a #10-24 × 1⅜" long machine screw and nut, as shown in Figure **G**.

A PAC, frame highlighted

B PAC, frame detail

C PAC, frame, sides assembled

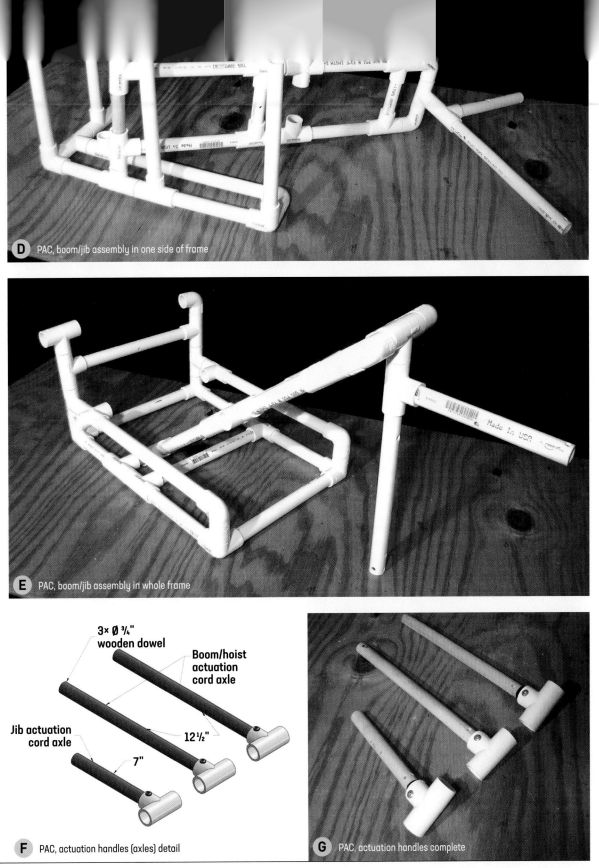

D PAC, boom/jib assembly in one side of frame

E PAC, boom/jib assembly in whole frame

3× Ø ¾"
wooden dowel

Boom/hoist
actuation
cord axle

Jib actuation
cord axle

12½"

7"

F PAC, actuation handles (axles) detail

G PAC, actuation handles complete

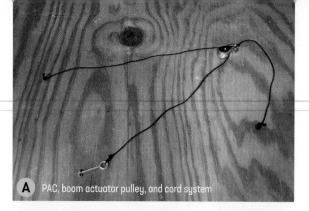

A PAC, boom actuator pulley, and cord system

B PAC, boom actuator, installing the fixed pulley

Staying on Track: Eye Bolt Eye Opening

It can take a fairly large amount of force to bend the eye of an eye bolt open. You will need to hold the threaded end securely using either a wrench or (preferably) a vise. To not damage the threads, wrap the threads with several layers of duct tape before clamping on to the threaded end. Once you have secured the threaded end, use vice grips to grab the eye end of the eye bolt near where the free end is bent around to. Apply force with the vice grips to open up the eye enough to get the pulley swivel loop over it. You now will have a hook. You can leave it this way or bend it back shut to be retained on the pulley. To close it back squeeze it shut with a pair of pliers.

Use duct tape to protect threads when opening an eye bolt

STEP 7

In this step, we are hooking up the boom actuation pulleys, cords and cord axle. This pulley system, as we are going to build it here, gives us an ideal mechanical advantage of 4. Start by making an assembly consisting of a single pulley, two lengths of cord, and an eye bolt. The long length of cord is 24" long, and the short cord is 12" long. Tie one end of the long cord to the eye bolt. Then feed the free end through the pulley as shown, and tie a slip knot to prevent the cord from coming back through. Tie one end of the short cord to the swivel loop of the pulley. Tie another slip knot in the free end of the short cord. We install this into the crane at a later step, but for now, just set it aside (Figure A).

Now you need to open up an eye bolt eye enough so that you can get the swivel loop of a second single pulley onto it, as shown in the *Staying on Track: Eye Bolt Eye Opening*. Install the pulley/eye bolt assembly into the 3/16" diameter hole you drilled in the lower cross member of the boom, and secure it using a nyloc nut. Do not fully tighten the eye bolt against the PVC cross member; leave it loose so that it can move or swivel slightly during operation (Figure B).

Next, going back to the cord, pulley, and eye bolt assembly we made previously, thread the short end of the cord through the pulley you just mounted to the boom. Insert a zip tie through the slip knot loop in the free end of the short cord as shown in Figure C. Loosely secure the zip tie around the frame just above the longitudinal frame member, and below the upper cross member of the frame as shown in Figure D. With the boom in its forward-most position (Figure D), pull the short cord in until the movable pulley just touches the pulley fixed to the boom (Figure E). The cord is pulled in by slowly pulling the free end of the zip tie, thereby reducing its loop length. This is how you can fine-tune the length of the cord.

It is now time to install the boom actuation cord axle constructed in step 6. Slide the axle in through the open tee at the top rear of the frame. Push the axle through the open fitting and into the closed receiver fitting on the other side, as far as it will go.

Each of the three actuation cord axles is secured in the frame or boom using a wood screw. Mark a point on the dowel approximately 3/16" in from the inside of the open fitting. Drill a pilot hole partway into the dowel and install

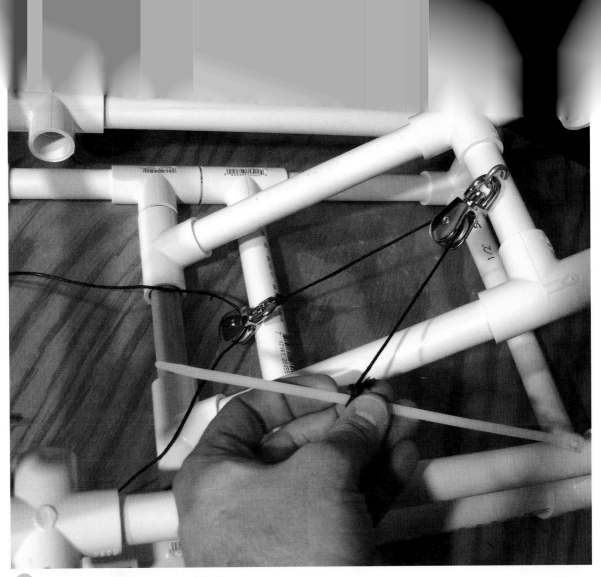

C PAC, boom actuator, securing the cord with a zip tie

D PAC, boom actuator, tightening the cord

E PAC, boom actuator, moveable and fixed pulleys touching

A PAC, boom actuator, installing the boom actuation axle

B PAC, boom actuator, attaching the cord to the axle

C PAC, boom actuator, axle, and cord/pulley system installed

a wood screw. Place the screw so that the axle can still rotate freely, but cannot shift longitudinally, along its axis, very much (Figure **A**). Install another wood screw toward the middle of the boom actuation axle, leaving some of the threads exposed (the exact placement of this screw is not that important). Take the free end of the long cord and slip the loop over the screw (Figure **B**). Pull the slip knot tight, and then carefully tighten the wood screw to secure the cord in place. Your assembly should now look like Figure **C**.

Now, wind the loose cord up on the boom actuation axle. Make sure that you wind it tightly by allowing the loose cord to slide through your fingers, providing some friction while you are winding the cord. Be careful that the cord stays on the pulley wheel while tightening. Once both the long and short cords are tight, the system is ready for action.

At this point you may be thinking, "What keeps the axle from rotating on its own when I let go of the handle?" That is an excellent question, with a pretty simple solution! Let's install a simple locking mechanism. While holding the axle still, cross-drill through the tee and wooden dowel using a $3/16$" diameter drill bit. Now, line up the hole in the dowel with the hole in the tee fitting and then drop a pin (or, in our case, a machine screw) through, thus locking the axle to the frame, as shown in Figure **D**. (Optional: To get a bit more precision in where you can lock the axles, rotate the axle by a quarter-turn, and carefully drill through the wooden dowel again using the hole you previously drilled through the PVC tee fitting as a guide.) Figures **E**, **F**, **G**, **H**, and **I** shows the boom actuated and "locked" in multiple positions.

Although not specified in later steps, you need to repeat these instructions to install locking mechanisms in each of the two remaining actuation cord axles.

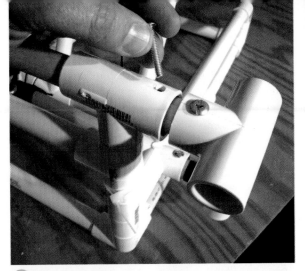

D PAC, installing axle "locks"

E PAC, boom in lower position (cords are extended)

F PAC, boom in lower position (cords are extended)

G PAC, boom in higher position (cords are retracted)

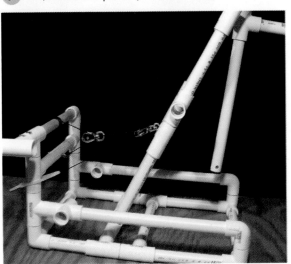

H PAC, boom in higher position (cords are retracted)

I PAC, boom in higher position (cords are retracted)

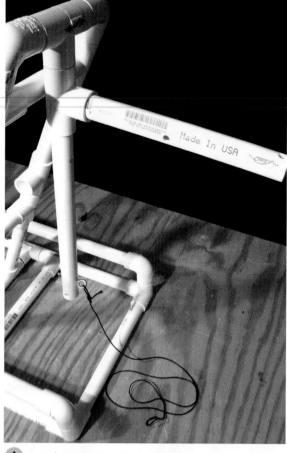

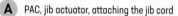

A PAC, jib actuator, attaching the jib cord

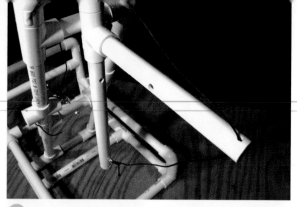

B PAC, jib actuator, installing the jib cord

C PAC, jib actuator, installing the jib actuation axle

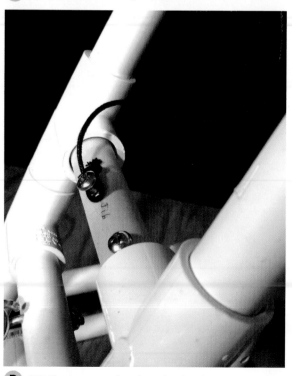

D PAC, jib actuator, attaching the cord to the axle

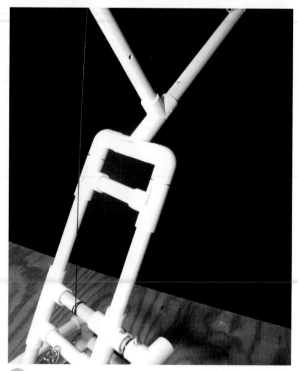

E PAC, jib extended

STEP 8

Now that the boom actuation is working, let's make the jib move. In this step, we hook up the jib actuation cord and cord axle system. Relative to the boom actuation system, this one is quite simple. Start by installing an eye bolt into the 3/16" diameter hole at the end of the jib. Now, tie one end of a length of cord 36" long to the eye bolt, as shown in Figure **A**. Feed the free end of the cord through the upper 3/16" diameter hole in the jib mast, as shown in Figure **B**. Insert and secure the jib actuation cord axle in the boom, as shown in Figure **C**. Fasten the cord to the axle using the same procedure that was presented in step 7 (Figure **D**). Now, drill the boom fitting and axle to receive a locking pin, as presented in step 7. Finally, wind the cord tightly on the axle until the jib begins to move. At this point, you can lock the axle, and the jib should remain at the angle you set it to relative to the boom. Figures **E** and **F** show the jib in different positions.

STEP 9

The last step in making your fully functional crane is to install the hoist cord, pulley, hoist cord axle, and hook system. Start by threading a 120" length of cord through the hoist system's single pulley, and tie a slip knot loop at each end of the cord, as shown in Figure **G**.

Next, secure a hook to the swivel loop of the pulley via a small length of cord or a zip tie. How you make your hook, or what you make the hook out of, is not that critical. We simply used what we had laying around. We used a wood screw hook. Wrap the top of your hook a few times with a bit of cord, and then tie it in a knot (Figure **H**). Further secure it by wrapping the cord with some duct tape. For tips on making a hook from coat hanger wire, see the *Staying on Track: Coat Hanger Wire Hook* below.

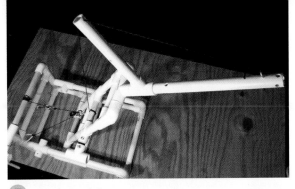

F PAC, jib extended

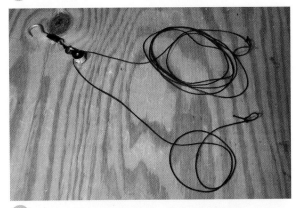

G PAC, hoist actuator, pulley and cord system with hook

H PAC, homemade hook

Staying on Track: Coat Hanger Wire Hook

Coat hanger wire also can be used to fashion a hook. Just cut a length of the coat hanger wire approximately 3" long. Then, using needle-nose pliers, bend one end into a tight, closed loop. Next, fashion the other end into a hook as shown in the figure to the right You can then tie a cord around the small, closed loop to attach it to the crane.

Fashion a hook from coat hanger wire

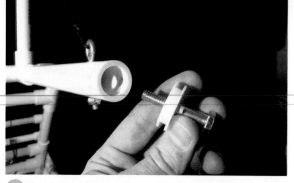

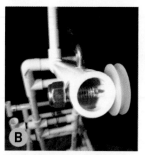

A PAC, hoist actuator, installing the jib end cord guide

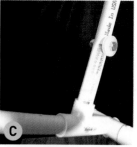

B PAC, hoist actuator, jib end cord guide installed

C PAC, hoist actuator, installing the jib mast cord guide

D PAC, hoist actuator, installing the hoist actuation axle

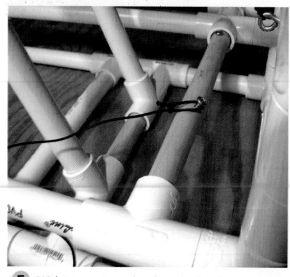

E PAC, hoist actuator, attaching the cord to the axle

The hoist cord is guided by two idler pulleys attached to the jib. One is mounted to the jib mast, while the other is installed at the end of the jib. The one at the end of the jib serves as the final guide for the hoist cord and hoist hook. The idler pulleys are made from repurposed, sliding-door guide wheels. The guides originally came with axles, but they are not used here (save them, as you may find a use for them later). In this project, we will be using ¼"-20 × 1½" bolts as the axles. The purchased door slide guide wheels fit a bit too snug on the bolts and do not allow the guides to spin freely. So, using a ¼" drill bit, drill out the axle hole of the door guide wheels so that they spin freely on the smooth shank of the bolt.

Install one of the guide pulleys at the front of the jib through the ¼" diameter hole drilled when you built the jib. Pushing on the bolt head, slide the pulley over until it makes contact with the PVC pipe (Figure **A**). Then secure the bolt with a nyloc nut, leaving a small gap (approximately ¹⁄₁₆" to ⅛") between the nut and the PVC pipe (Figure **B**). Later, you will see that this gap serves an important role in the hoist system. Mount the second idler pulley to the jib mast again using a nyloc nut through the ¼" diameter hole toward the middle of the mast (Figure **C**). Make sure not to tighten it too much; the pulley needs to spin freely.

Now, install the hoist cord axle. Referring to the instructions in step 7, slide the axle into the frame and secure it in place using a wood screw, as done with the other two axles. Also, cross-drill the frame tee and axle dowel to make the axle lock (Figure **D**). Add another wood screw to the axle, as shown in Figure **E**, and loop one end of the hoist cord over it. Tighten the screw to secure the cord to the axle. Take the other end of the hoist cord and secure its' looped end over the threads between the nut and PVC pipe of the idler pulley axle at the end of the jib (Figures **F** and **G**). Tightly wind some of the hoist cord onto the hoist cord axle (Figure **H**). Finally, place the hoist cord in the groove of both of the hoist idler guide pulleys, as shown in Figure **I**.

The crane is now complete!

While utilizing your new crane, you may notice that there is a limit to how much weight you can lift before the crane begins to tip over. This limit will vary with respect to the position of the boom and jib of your crane. The further out you lift the load from the base (Figure **J**), the less weight you can lift. In the configuration shown in Figure J, you can lift up to about half a pound before the crane begins to tip over. To lift a larger amount of weight, you must orient the boom and jib such that it moves the lifting point closer to the base. The crane can lift roughly two pounds in the configuration shown in Figure **K** before tipping. Consider what you learned in Chapter 6 - Levers to think through why the orientation of the boom and jib (i.e., lifting point) changes how much the crane can lift without tipping over. To increase the amount of weight your crane can lift, add some weight to the base. The further away from the lifting point that you add this weight, the less amount of weight it will take to counterbalance the load (the left side of the base as shown in Figures J and K.) Full-sized cranes use large counterweights in their bases to increase the amount of weight they can lift.

Pulleys are versatile, simple machines that are found in many different applications around us. They are used in the obvious ways, like in cranes that lift heavy objects, and are also common in nearly every type of vehicle, from automobiles to airplanes, and nearly all machines. We hope that you feel more informed about these useful simple machines, and will find many ways to incorporate them into your maker projects. ❂

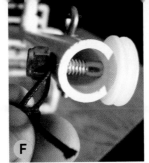

F PAC, hoist actuator, attaching the cord to jib end

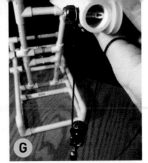

G PAC, hoist actuator, installing the hoist pulley

H PAC, hoist actuator, tightening the hoist cord

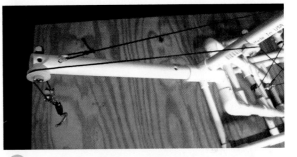

I PAC, hoist actuator, threading the hoist cord over the guides

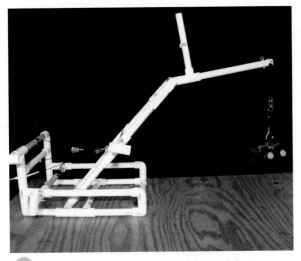

J PAC, lifting a small load with boom and jib extended

K PAC, lifting a large load with boom and jib oriented to move the load closer to the base

8

Gears and Gear Trains – Grinding, Isn't It

In this chapter, we are going to explore gears: fantastically simple machines that can be found in nearly every moving mechanism. Gears are wheels with teeth that mesh and work together to transmit rotational energy from one shaft to another. The teeth engage each other mechanically, ensuring that there is no slip from one gear to another. Makers use this engagement to vary speed and change direction.

Per our definition of a simple machine (a mechanism that changes speed and/or direction), gears are a simple machine. Like the other simple machines that you have already been introduced to in this book, gears have been around for ages. The earliest known example of a geared mechanism was discovered in 1902 by sponge divers near the Greek island of Antikythera. It was a bronze mechanism comprised of 37 individual gears, which was used to calculate the phases of the moon, dates of lunar and solar eclipses, and the positions of constellations of the zodiac. This contraption, dubbed the "Antikythera Mechanism," has been dated to around the 2nd century BC. It might be the first analog computer in gear form!

Adobe Stock - royyimzy

The Antikythera Mechanism displayed at the National Archaeological Museum in Athens, Greece

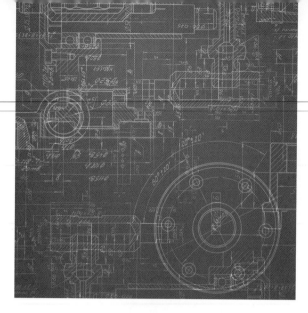

Types of Gears

Gears come in a wide variety of shapes and sizes, and are used in nearly every mechanical system. In this chapter, we introduce you to some of the most common gear types that you may come into contact with as a maker.

- **Spur Gears** are the simplest of gears, and are what most people envision when thinking of gears. They are wheels or cylinders with teeth arranged radially outward from the center (Figure **A**). When working as a system (e.g. with multiple spur gears meshed together), the shafts of spur gears must be parallel to each other. Spur gears work well at lower speeds, but noisy at high speeds because of the gear teeth impacting each other as they engage.

- **Helical Gears** are wheels or cylinders with the teeth cut at an angle relative to the axis of the gear (Figure **B**). The teeth are curved along a helix shape (hence the name helical gear; they are sometimes called herringbone gears). Due to this helical profile, they engage more gradually than spur gears, thus providing smoother and quieter running gears. Helical gears can be arranged with their shafts parallel or at various angles relative to each other, depending on the teeth configuration. When in operation, helical gears tend to have higher friction due to the "sliding" of the gear teeth against each other. Also, helical gears have an axial (or thrust) load on the gear shaft, since the gear teeth engage along a curve instead of straight on; in other words, the gear tends to "screw in or out" as it rotates. Double helix gears eliminate this issue by having a double set of teeth oriented in opposite directions. The axial thrust developed by one side of the gear is negated by the axial thrust generated by the other side in the opposite direction. Double helix gears are difficult to make and are very expensive, but are necessary in certain instances.

A Spur gear set

B Helical gear set

- **Bevel Gears** are wheels or cylinders with teeth arranged around a cylindrical cone with the tip lopped off. When two bevel gears with the same diameter and number of teeth are working together in a system, they are known as miter gears. Miter gears only change the gear shaft axis angle or axis of rotation; there is no change in rotational speed relative to the input or output gear. Similar to spur gears, bevel gears run well at lower speeds, but they get noisy at higher speeds.

Engineers have combatted this issue by developing a bevel gear with a curved tooth profile. This type of gear is known as a spiral bevel gear (Figure **C**).

- **Hypoid Gears** look very similar to spiral bevel gears, but they are designed so that the axis of the gear does not intersect the axis of the mating gear (Figure **D**). The curved teeth are shaped along a hyperbola (hence the name hypoid gear). Hypoid gears produce much less noise than other gears types and also run more smoothly. They almost always have an axis of rotation 90° to that of the mating gear, and since the shaft axes of both gears do not intersect, it is possible to support both ends of the shafts of both gears. Hypoid gears are commonly used in automotive axles, where a gear reduction and right-angle change of direction are required. The disadvantage to hypoid gears is that their curved tooth profile creates an axial thrust force (similar to helical gears) that must be handled by support bearings. Also, they tend to be relatively expensive.

- A **Worm Gear** set is comprised of two gears: a worm and a worm wheel. In general terms, a worm gear looks like a screw and a worm wheel resembles a spur gear. A worm can have a few teeth, or one long tooth that is wrapped continuously around its base like a screw (Figure **E**). Worm gears are typically used when a large gear reduction is required. The gear ratio of a worm drive is simply the number of teeth of the worm wheel divided by the number of teeth of the worm. So, for example, if a worm wheel has 40 teeth, and the worm has only one, then the worm has to rotate 40 times for every one revolution of the worm wheel (40:1), producing a 40:1 gear reduction ratio. Due to the large gear ratios of worm gear sets, they can also transmit a very large amount of torque. For example, when set in an ideal, frictionless state, our 40:1 ratio worm gear provides a torque 40 times greater than the input torque to the worm. Finally, worm gear sets typically cannot be back-driven, meaning that the worm wheel cannot drive the worm when a torque is applied to it. The worm gear set can only be moved by rotating the worm.

A winch is a great application of a worm gear set. The high gear ratio and subsequent increase in torque in only two gears makes this an excellent gear choice for a winch. Also, since the worm gear set cannot be back-

C Spiral bevel gear

D Hypoid gear set

E Worm gear set

driven, the winch's load does not fall when no torque is applied to the crank handle (or worm). It is important to note that a worm bears an axial load directly related to the load on the worm wheel. The bearings used to support a worm must be able to handle this axial or thrust loading along with the rotation of the worm.

- **Rack and Pinion Set:** So far, all of the gear types presented transmit only radial or rotational motion. A rack and pinion gear set, however, is used to convert rotational motion into linear motion. The pinion is a gear that looks like a regular spur gear. The rack is a straight bar with gear teeth cut along one edge of the bar (Figure **A**). In a rack and pinion system, the rack is constrained such that it can only translate back and forth along a linear axis. The pinion gear engages the rack, and when it is rotated, the rack is driven in one linear direction. The rack can be driven in the opposite direction by rotating the pinion in the opposite direction. Car steering uses a rack and pinion gear set to translate the rotational torque from the steering wheel to a linear force applied to the tie rods, causing the wheels to turn.

- **Planetary Gear Sets** (also known as epicyclic gear trains) are comprised of four main elements. The first is an outer ring gear with inward-facing teeth. Engaged with the ring gear are multiple planet gears. These planet gears are linked together via a carrier. A single, central gear known as the sun gear is engaged with all of the planet gears (Figure **B**). The ring gear, carrier and sun gear rotate about the same axis. Each planet gear rotates on an axis constrained by the carrier. As the carrier rotates, the planet gears both rotate about their individual axis while simultaneously "orbiting" about the sun gear. Planetary gear sets are used to transmit a large amount of torque in a relatively small, compact package. They are also able to behave differently, depending upon which part of the gear train is held stationary. They can work as a reduction gear, an increasing gear, or even a reverse gear. This makes planetary gears well suited to applications like automatic transmissions. In fact, this type of gear set is used in many applications ranging from heavy construction equipment to bicycle gear hubs.

A   Rack and Pinion

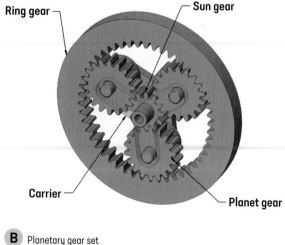

Ring gear

Sun gear

Carrier

Planet gear

B   Planetary gear set

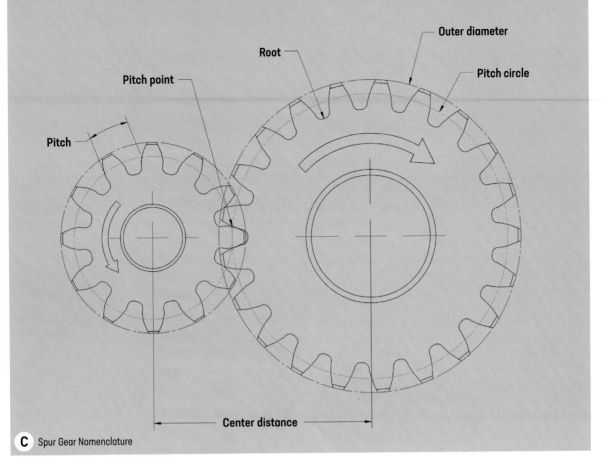

Pitch point

Root

Outer diameter

Pitch circle

Pitch

Center distance

C Spur Gear Nomenclature

General Gear Nomenclature –
The Simplest Example (the Spur Gear)

Before we get into any gear train design specifics and analysis, we need to discuss some general gear terminology. To accomplish this, let's look at our simplest and perhaps most common gear example: the spur gear (Figure **C**).

- **Root:** This is the bottom-most part of the gear tooth. The tooth height of two engaging gears is cut such that it is slightly less than the root of the gear. This is so that the tip of each tooth does not hit the root of the gear with which it is engaged.

- **Outer Diameter:** This is a circle describing the outermost extent of a gear. The outer diameter circle lies on the tip of each tooth.

- **Pitch:** The pitch of a gear is the distance between the same point on one tooth and that of the next.

- **Pitch Circle:** The pitch circle of a gear runs roughly through the center of the gear teeth. When properly aligned, the pitch circles of two engaged gears touch at a single tangent point.

- **Pitch Point:** The point where the two pitch circles of two engaged gears touch. This is the point where the gear teeth make contact.

- **Center Distance:** This is a critical part of gear system design that has to be correct for the gears to engage properly. It is the distance between the axis of rotation of one gear to that of a second gear. To find the proper center distance between two gears, add the pitch circle diameter of both gears and then divide by two.

Gear Ratios

When designing a mechanical system that uses gears to transmit power, it is important to understand what the output of your system is doing relative to the input of the system. This relationship is defined by something known in mechanical engineering as a **gear ratio**. The overall gear ratio of a system is defined by the ratio of the rotation of the output shaft to the rotation of the input shaft.

Because there is no slip between gears, they have "exact velocity ratios." What does that mean? For example, if one gear driving another gear rotates at 100rpm (revolutions per minute), and the driven gear rotates at 200rpm, the driven gear will always rotate at twice the rpms of the driver gear. This makes the gear ratio 1:2. If the rotational speed of the driver gear increases to 400rpm, the driven gear will rotate at 800rpm. Additionally, since there is no slip between gears, the linear speed at the pitch circles of two gears must be the same. This is how we can define and begin to understand gear ratios. If a larger gear is engaged with a smaller gear, then the rotational speed of the smaller must be greater than that of the larger gear in proportion to their pitch diameters. The pitch circle of the larger gear has a greater circumference than that of the small gear. Given that there is no slip between the gears, and the linear speed of both pitch circles must be equal, then for every revolution of the bigger gear the smaller gear must rotate more revolutions to equal the circumference of the larger gear.

A gear ratio can be greater than 1, equal to 1 or less than 1. When gears are used to enable the use of a higher rpm, lower torque motor to drive a slower rpm output requiring a higher torque, it results in a gear ratio of greater than 1. This greater-than-1 scenario is very common in vehicles that require high amounts of torque, such as trucks and tractors. A gear ratio of less than 1 gives you the opposite result: the output shaft of the system rotates at a greater rpm than the motor or input shaft rpm, at the expense of torque.

A gear ratio equal to 1 means that the input rpm is equal to the output rpm. This is similar in concept to having a mechanical advantage of 1. You may recall from our chapters on levers and pulleys that a system with a mechanical advantage of 1 is only useful if a change in the direction of the input force relative to the output force is required. There is no reduction or increase in force.

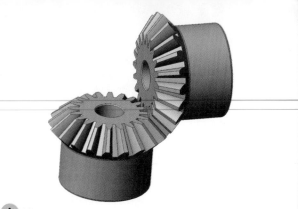

A Miter gear set

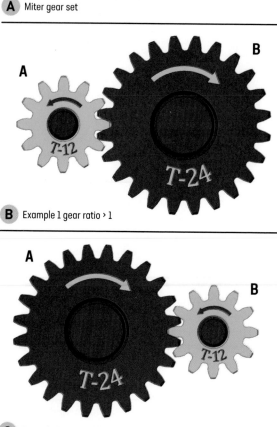

B Example 1 gear ratio > 1

C Example 2 gear ratio < 1

D Example 3 Fear Eatio = 1

Sometimes use a gear ratio of 1 is used to change the direction of rotation of the output relative to the input without increasing or decreasing the relative rpms of the gear shafts. In addition to changing the direction of rotation, you may also need a gear system with the gear ratio of 1 to allow for the input shaft to not be parallel to the output shaft. A good example of this is a bevel gear set with both gears having the same number of teeth. As previously mentioned, this type of bevel gear set is known as miter gears. Each gear's axis is oriented at 90° to the other. The input to one of the gears results in the other gear rotating at the same rpm, but its shaft is oriented perpendicular to that of the first gear's shaft. Also, the second gear rotates in the opposite direction to the first (Figure **A**).

To help you better understand the concept of gear ratios, let's look at a few basic example calculations. In example 1, we look at a two-gear system where a smaller gear (A) is driving a larger gear (B), as shown in Figure **B**. As we've said before, the gear that is driving is known as the "driver", and the gear that is being driven is simply referred to as "driven."

From earlier in our discussion, we know that the gear ratio is calculated by dividing the rotational velocity of the output shaft by the rotational velocity of the input shaft. But how do we directly relate this to gears? Knowing that there is no slip between the two gears, and that the linear speed of the pitch circles must be the same, we can base our calculations on the relative number of teeth on each gear. In our current example, gear B has 24 teeth and is the driven gear. Gear A is the driver, and it has 12 teeth. The gear ratio is calculated as follows:

Gear Ratio $\rightarrow \left(\dfrac{T_{Driven}}{T_{Driver}}\right) = \left(\dfrac{T_B}{T_A}\right) = \left(\dfrac{24}{12}\right) = \left(\dfrac{2}{1}\right) = 2 \text{ or } 2:1$

The gear ratio for this example is calculated to be a value of 2. As previously noted, a gear ratio greater than 1 means that the driver, or input gear, must rotate faster than the output, or driven gear. In this case, the smaller input gear must rotate 2 revolutions for every 1 revolution of the larger output gear. With this in mind, we can write this gear ratio as 2:1.

When talking about the rotational speed of gearing, it is common to refer to a gear ratio as a velocity ratio. Let's say that gear A is rotating at 20rpm. In mechanical engineering calculations, rotational speed is commonly denoted by the Greek letter ω. How do we calculate how fast gear B will rotate? We divide the rotational speed of the driver (ω_{Driver}) by the gear ratio (velocity ratio). Gear B will rotate at half the rpm as gear A, or 10rpm.

$\left(\dfrac{\omega_{Driver}}{Gear\ Ratio}\right) = \left(\dfrac{20\ rmp}{2}\right) = 10\ rpm$

In example 2 (Figure **C**), we look at a scenario where the driver gear (A) is larger than the driven gear (B). We should expect to see that the driven gear will rotate faster than the driver gear (equation below). In this case, for every one rotation of the larger input (driver) gear, the smaller output (driven) gear will rotate twice. With this in mind, we can write this gear ratio as 1:2, which is a gear ratio of less than one.

Gear Ratio $\rightarrow \left(\dfrac{T_{Driven}}{T_{Driver}}\right) = \left(\dfrac{T_B}{T_A}\right) = \left(\dfrac{12}{24}\right) = \left(\dfrac{1}{2}\right) = 0.5 \text{ or } 1:2$

A gear ratio of less than one means that the driven gear rotates faster than the driver gear. If the driver gear is rotating at 20rpm then the rotational velocity of the driven gear is as follows:

$\left(\dfrac{\omega_{Driver}}{Gear\ Ratio}\right) = \left(\dfrac{20\ rmp}{0.5}\right) = 40\ rpm$

For example 3 (Figure **D**), we look at two gears with the same number of teeth meshed together. Why would you want to have this sort of gearing scenario? In some instances, gears are used to convey rotational motion from one shaft to another shaft without changing the rotational speed of the output shaft. This can easily be achieved by linking the two shafts by gears with the same number of teeth on each shaft. But what happens to the direction of rotation of the output shaft? It is reversed, just as with all other spur gear arrangements. If the input shaft is rotating clockwise, then the output shaft rotates counter-clockwise. This reversal of direction of the driver gear relative to the driven gear also occurs in both examples 1 and 2. However, in the case of the first two examples, there is also a rotational speed change.

Gear Ratio $\rightarrow \left(\dfrac{T_{Driven}}{T_{Driver}}\right) = \left(\dfrac{T_B}{T_A}\right) = \left(\dfrac{21}{21}\right) = \left(\dfrac{1}{1}\right) = 1 \text{ or } 1:1$

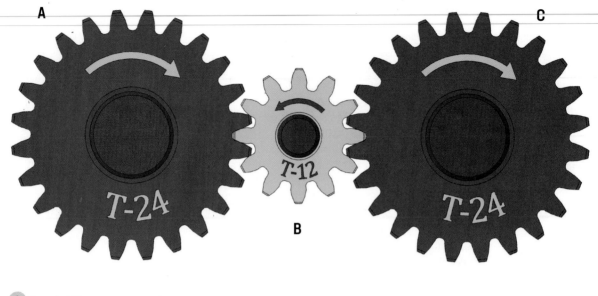

A Example 4 idler gear, gear ratio = 1

Once again, if the driver gear (A) in this example is rotating at 20rpm, we can show that the driven gear (B) is rotating at the same rate, meaning that the gear ratio is 1:1, or equal.

$$\left(\frac{\omega_{Driver}}{Gear\ Ratio}\right) = \left(\frac{20\ rmp}{1}\right) = 20\ rpm$$

Up until this point, all of our gear ratio examples have only had two gears involved. In our next example (example 4) we introduce a third gear. Does the gear in the middle (B) change the gear ratio between gears A and C? The answer is, no. In this example, gear B is there to act as what is known as an **idler gear** (Figure **A**). Its' only purpose is to enable gear C to rotate in the same direction as gear A. Gear B does not influence the overall gear ratio at all.

To calculate the gear ratio for this example, we can approach it in the same way as all of the previous examples. We simply calculate the ratio of the number of teeth of the final driven gear (C) divided by the number of teeth of the driver gear (A).

Gear Ratio $\rightarrow \left(\frac{T_{Driven}}{T_{Driver}}\right) = \left(\frac{T_C}{T_A}\right) = \left(\frac{24}{24}\right) = \left(\frac{1}{1}\right) = 1\ or\ 1:1$

Now, let's show mathematically that the idler gear does not affect the overall gear ratio. To do this, we look at the first two and second two gears separately then combine them into a system equation. Let us explain. Gear A is the driver with respect to gear B, which is the driven gear. So, start with this gear ratio:

Gear Ratio B/A $\rightarrow \left(\frac{T_{Driven}}{T_{Driver}}\right) = \left(\frac{T_B}{T_A}\right) = \left(\frac{12}{24}\right) = \left(\frac{1}{2}\right) = 0.5\ or\ 1:2$

We now examine gears B and C. Gear B is the driver relative to gear C. The gear ratio for these two gears is as follows:

Gear Ratio C/B $\rightarrow \left(\frac{T_{Driven}}{T_{Driver}}\right) = \left(\frac{T_C}{T_B}\right) = \left(\frac{24}{12}\right) = \left(\frac{2}{1}\right) = 2\ or\ 2:1$

Now we need to look at the overall system, the gear train as a whole. To calculate the overall gear ratio we multiply the gear ratio between gears A and B by the gear ratio between gears B and C.

Overall Gear Ratio $\rightarrow$
$$\left(\frac{T_{Driven}}{T_{Driver}}\right) = \left(\frac{T_B}{T_A}\right)\left(\frac{T_C}{T_B}\right) = \left(\frac{12}{24}\right)\left(\frac{24}{12}\right) = \left(\frac{24}{24}\right) = 1\ or\ 1:1$$

Note that T_B appears in both the numerator and denominator of the overall equation. Therefore, T_B is canceled out, illustrating how gear B does not affect the overall gear ratio.

The Ordinary Can Opener –
A Gear Ratio Discussion and Intro to Compound Gears

Gears are parts of many everyday devices. Take, for instance, the ordinary can opener. A can opener requires a very large amount of torque to positively grip the can and rotate it at the same time a blade tears the metal lid away from the body of the can. Fortunately, the action of opening a can is typically done at a fairly slow speed. To achieve a large torque where the can is gripped and driven without having a very large, expensive motor, an electric can opener uses a gear train.

Have you ever noticed, while running your electric can opener, how it makes a very mechanical, almost growling-type sound? What you're hearing is the gear train reducing the high speed of the electric motor, while increasing the torque and lowering the rotational speed at the output of the can opener. Figure **B** shows an old, under-counter style can opener. If we take the rear cover off, we can see the motor rotor (the part of the motor that spins) and a simple gear train (Figure **C**).

Figure **D** shows a 3D representation of the can opener gear train. What do all of these gears accomplish? A gear train like this is known as a **compound gear set**. In a nutshell, this gear train provides a two-stage gear reduction from the motor input to the can opener output, or can drive. The yellow gear (gear A) represents the drive or input gear. This is the gear that is attached to the motor shaft. Look closely at the drive gear: its teeth have a spiral look to them. This is an example of a spiral gear. A spiral gear, in this application, allows for more tooth contact with the plastic gear that it is engaged with, thus distributing the tooth loading over a larger area.

The first reduction stage happens between gear A's eight teeth and the 60 teeth of the orange mating gear (gear B). This represents a gear ratio of 7.5: 1, a significant reduction. The green gear (gear C) is fixed to gear B such that they must rotate together. Note that even though gear B and gear C have a different number of teeth, this does *not* change the gear ratio; gear C still makes only one rotation for every rotation gear B makes. Finally, the blue gear (gear D) represents the larger spur gear that is fixed to the output drive of the can opener. Its 52 teeth are driven by the eight teeth of gear C, resulting in the second gear reduction of 6.5:1.

B   Under-counter Can opener

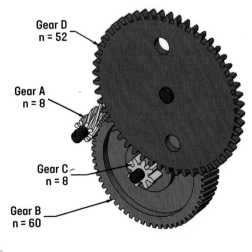

C   Can opener with back over removed

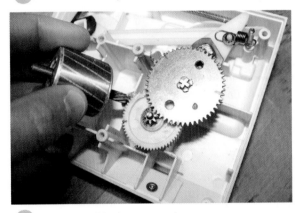

Gear D
n = 52

Gear A
n = 8

Gear C
n = 8

Gear B
n = 60

D   Can opener gear train (compound gears)

Now, let's talk about what is going on in this gear train in a mathematical sense, and calculate the overall gear ratio for this gear train. The gear ratio is a single number that represents the number of rotations the motor needs to turn for every single rotation of the output of our can opener. It also represents the increase in torque between the motor and the output drive of the can opener. This increase in torque is calculated by multiplying the motor input torque by the gear ratio. As we have learned with other mechanical systems discussed in this book, the previous statements assume a perfect or ideal system, even though all real systems waste some energy in friction or heat. This is also true for gear trains.

The next equation shows the gear ratio calculation for this compound gear train. We start analyzing this gear train by initially looking at the driven, final blue gear (D). This gear is driven by the small green spur gear (C), which acts as a driver for gear (D). The first part of our equation represents the second stage of the gear train's overall gear ratio. The first stage is comprised of the orange gear (B) as the driven gear and the yellow gear (A) as the driver. This gear ratio stage appears in the second part of the equation. Combining these two gear ratio stages, we get the overall gear ratio of our compound gear train.

$$\left(\frac{T_{Driven}}{T_{Driver}}\right) = \left(\frac{T_D}{T_C}\right)\left(\frac{T_B}{T_A}\right) = \left(\frac{52}{8}\right)\left(\frac{60}{8}\right) = \left(\frac{3120}{64}\right) = 48.75 \text{ or } 48.75 : 1$$

Note also that we can get the same result by multiplying together the two individual gear ratios of the compound set.

$$R_{TOTAL} = R_{AB} * R_{CD} = 7.5 * 6.5 = 48.75$$

Now that is a gear reduction! The can opener motor spins 48.75 revolutions for each one revolution of the final drive or output of the can opener.

Gear-Based Device Discussion: Differential

In this section, we discuss a really cool, gear-based device known as a differential (Figure **A**). A differential can be found in pretty much any motor vehicle with wheels and a single engine. A differential serves as the final link in the powertrain of a vehicle just before the wheels. Take for instance a typical rear-wheel drive car. The power to propel the car is supplied by an engine which is directly coupled to a gearbox or transmission. The transmission allows for gearing changes that enable the engine to transmit more torque at the expense of speed (lower gears) or more speed at the expense of torque (higher gears). This is transmitted through a driveshaft to a final fixed ratio gearbox known as a differential (Figure **B**).

The output of the transmission is the input to the differential. The differential serves three primary purposes in the drivetrain of an automobile. First, it provides a way to redirect or change the direction of the axis of rotation of the engine, transmission, and driveshaft to that of the axles driving the wheels. In the case of a rear-wheel drive car, the axis of rotation of each of the axles is 90° to that of the rest of the drivetrain. The second purpose of a differential is to provide a final fixed gear reduction, thereby slowing the output of the transmission while increasing the amount of torque available to the wheels. The third, and perhaps most notable, purpose of the differential is to provide a means by which to supply torque to each of the two drive wheels, while simultaneously allowing them to rotate at different speeds. This is where the differential gets its name.

So, why is it necessary to allow the two drive wheels of an automobile to be able to rotate at different speeds while being driven? This can be answered through simple geometry. As a car turns around a corner, the inner wheels are going through a smaller radius arc relative to that of the outer wheels (Figure **C**). So, if we assume that the wheels can rotate freely and are therefore not slipping, then as the car goes through the corner the outside wheels are required to go a slightly further distance than the wheels on the inside of the corner. If in fact there is no slip, then the outside wheel must rotate faster than the inside wheels through the corner. This is what the differential enables the drivetrain of an automobile to do. If the differential was not there and both drive wheels were made to rotate at the same speeds, it would make

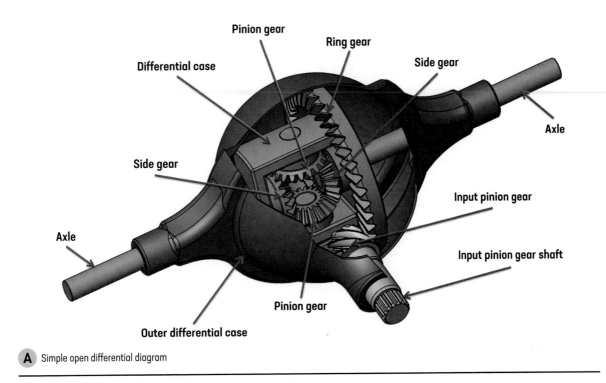

Pinion gear

Ring gear

Side gear

Differential case

Axle

Side gear

Input pinion gear

Axle

Input pinion gear shaft

Pinion gear

Outer differential case

A Simple open differential diagram

the vehicle very difficult to steer around corners. One wheel would have to skid or break traction in order to turn. Breaking traction and then regaining traction after the corner would also put a lot of shock load on the rest of the drivetrain, thereby reducing its lifespan.

In an "open" differential, the torque applied to each axle is always the same, regardless of the rotational speed of the axles relative to one another. As shown in Figure A, the drive torque from the driveshaft is applied to the input pinion (a small gear that is designed to mesh with a larger gear in a system). The input pinion then meshes with and drives the ring gear, which is rigidly attached to the differential case, causing the ring gear and case to rotate.

When both axle shafts are rotating at the same speed, then the gears within the differential case do not spin relative to each other. So, what happens if the axles are required to be driven at different speeds relative to each other, such as while a vehicle is making a turn, as shown in Figure C?

The differential case is still rotating as the ring gear is driven by the input pinion. However, now the gears within the differential case are also rotating relative to each other. This relative rotation of the pinion gears and side

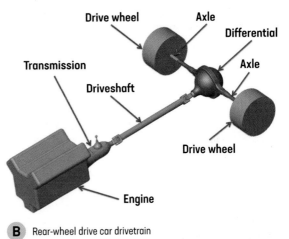

Drive wheel **Axle**

Differential

Transmission

Axle

Driveshaft

Drive wheel

Engine

B Rear-wheel drive car drivetrain

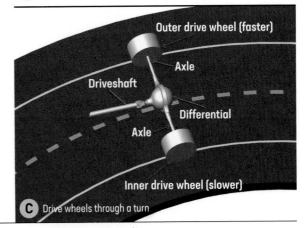

Outer drive wheel (faster)

Axle

Driveshaft

Differential

Axle

Inner drive wheel (slower)

C Drive wheels through a turn

A Brian's Drift Trike Industrial that he built from Alan Baum's plans (baumbuilds.com)

gears within the differential case is what allows for the axle/wheel on the inner side of a turn to go slower than the axle/wheel on the outer side, even though the torque that the differential supplies to both the inner and outer wheels remains the same for each, regardless of the different speeds.

If both drive wheels are driving without slipping (i.e., breaking traction) relative to the road, then the maximum torque available to the wheels while the vehicle is going straight or negotiating a turn is determined by the maximum amount of torque the engine and transmission can output. In situations where there is very low traction, and the wheel that is being driven by an open differential slips while the other wheel remains in traction (like on a patch of ice), what determines the torque available to the differential and both wheels? The torque available is the greatly reduced torque of the slipping or lower traction wheel. This is where an open differential can be a problem. If one of the drive wheels is unable to achieve

enough traction to allow for enough torque to propel the car, then the car will not be able to move at all. The slipping wheel will spin while the wheel with traction remains stationary.

An alternative to an open differential is a "limited-slip" differential. We aren't going into great detail about this type of differential, but let's take a brief look at it. In a limited-slip differential, there are special gear trains, or "clutches," within the differential. These devices are attached to the side gear/main axles of the differential such that they apply some resistance torque to the overall differential gear system. This added frictional torque becomes the limiting torque that the differential can deliver in a situation where one wheel has lost traction. This torque is tunable, and much greater than the torque available to an open differential when a wheel has completely lost traction.

There are times that a differential is not used at all. Take, for instance, Brian's Drift Trike, as shown in Figure A.

In the Drift Trike, the rear drive wheels are both driven at the same speed through a common, live-shaft axle. A live-shaft axle is a solid round bar on which the drive wheels are attached. The rubber tires of the trike are covered by a plastic pipe sleeve to reduce friction between the wheels and the ground. Without this reduction in friction, the Drift Trike would be un-drivable. But, for this vehicle, the relative slip of the inner and outer rear drive wheels and their collective slip relative to the ground is what makes this vehicle fun and exciting to drive. When Brian tested it initially without the sleeves, he found it almost impossible to steer as he headed straight for a tree! But, it wouldn't be a "drift" trike if it had a differential. The Drift Trike purposely spends a great deal of time with the wheels spinning relative to the road, in a broken traction state. The Drift Trike would effectively go nowhere if an open differential were used. One wheel would always be spinning while the other would not rotate at all. By eliminating the differential, the Drift Trike can supply full torque to each of the two drive wheels regardless of the amount of traction each one has.

In the following project, you build your very own open differential. Through building and experimenting with this project, you can gain a much deeper understanding of how this nifty little gear-based device works.

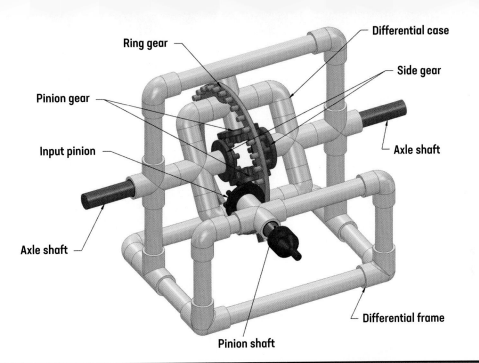

Ring gear

Differential case

Side gear

Pinion gear

Input pinion

Axle shaft

Axle shaft

Differential frame

Pinion shaft

After completing the Cog Gear Differential (CGD) project, you will have a fully functional open differential. This project can help you understand and appreciate functional gear systems, and systems. We hope that it guides and inspires you to use gears in your projects going forward.

WHAT MAKES UP THE COG GEAR DIFFERENTIAL (CGD)

A cog gear is a gear that uses pins to mesh with and drive pins on another cog gear, rather than conventional gear teeth. We use cog gears for this project because they are much easier to make, especially if you are making them without the aid of a 3D printer. Also, cog gears are much more forgiving of misalignment than bevel and hypoid gears, as would be used in a "real" differential.

The Cog Gear Differential is made up of three main modules or sub-assemblies: the central differential assembly, the differential frame, and the input pinion/shaft.

Materials:

» **PVC Pipe – ½" × 5' – standard schedule 40** (see cut list on the following page for details), **2 lengths** (MMC* part # 48925K91)

» **PVC Fitting – Elbow (90°) – ½", qty 10** (MMC part # 4880K21)

» **PVC Fitting – Cross – ½", qty 7** (MMC part # 4880K241)

» **PVC Fitting – Tee – ½", qty 2** (MMC part # 4880K41)

» **PVC Fitting – Side Outlet Elbow – ½", qty 2** (MMC part # 4880K631)

» **¾" Oak (Hardwood) Dowel** (see cut list for details), **qty 1 - 3' length** (MMC part # 96825K19)

Tools:

» **Hot Glue Gun and Glue**

» **Saw** (either handsaw, miter saw or PVC pipe cutting tool)

» **Pliers**

» **Drill**

» **³⁄₁₆" and ¼" Drill Bits**

» **Rubber Mallet**

* **MMC** = McMaster-Carr www.mcmaster.com

Cut List

Qty	Material	Cut Length	Notes
10	½" PVC pipe	1³⁄₈"	
1	½" PVC pipe	2"	
8	½" PVC pipe	4"	
1	½" PVC pipe	5⁵⁄₈"	
3	½" PVC pipe	8¹³⁄₁₆"	
2	Hardwood dowel	3½"	
1	Hardwood dowel	4½"	
2	Hardwood dowel	8"	
4	8-tooth spider cog		3D printed or cut from ¼" plywood and ¼" dowels
1	34-tooth ring cog		3D printed or cut from ¼" plywood and ¼" dowels
10	Axle spacers	Optional	3D printed (tapered)
1	Drill-drive adaptor	Optional	3D printed, #8 × 1½" socket head cap screw with nyloc nut

The central differential assembly is the heart of this differential. It houses and provides bearings for the two pinion gears, two side gears, and axles. This module also supports the large ring gear that ultimately drives the assembly.

The differential frame provides support and bearing surfaces for the central differential assembly via the axle shafts. The differential frame also supports the input pinion and shaft.

The input pinion and shaft is the simplest of all three modules, but it is equally as important. The input pinion and shaft transmit all of the drive torque to the CGD.

CONSTRUCTION

Make sure to prepare your cogs and other materials before starting construction.

If you have a 3D printer, download the print files provided on the website at gitlab.com/MakerMedia/ mechengformakers/tree/master, and print the five spider cogs and one ring cog, as shown in Figure Ⓐ.

Also, print the eight spacers to take up the "sloppiness" between the dowel axles and the pipe fittings in the frame. See *Staying on Track: Optional 3D Printed Spacers* on the following page for more information on why to use these and how to properly install them.

If you don't have a 3D printer available, you can create the cogs using ¼" thick plywood and ¼" diameter dowels. This process is described in *Staying on Track: Creating Cogs from Plywood and Dowels* on the following page. If you do not have a 3D printer and are using the plywood versions of the cogs, you can still build the CGD without the spacers. Your axles will be a bit wobbly, but it will still work just fine! As an alternative to using the 3D printed spacers to reduce this sloppiness, you can simply wrap a few layers of plastic packing tape around your axles. This will effectively increase the diameter of your axles, making then fit the PVC "bearings" more closely.

3D printed cogs for differential

A

Staying on Track: Optional 3D Printed Spacers

The ½" pipe fittings that hold the axles in place have through-openings that are more than ¾" diameter, which allows the ¾" dowels extra space to move around. This extra space allows the axle rotation to be "sloppy," so the 3D printed spacers are designed to fill the space and take up that sloppiness. The axles can function without these spacers, but will be wobbly within their pivots.

The spacers are designed so that the interior is straight to fit well on the dowels, but the outside is tapered to fit the tapered holes of the PVC fittings. Having the spacer in the correct orientation BEFORE sliding it onto the dowel is very important. To test the orientation before sliding it onto the dowel, slide it into the PVC fitting. If the spacer slides in easily and goes deep into the fitting, as seen in the image here, then it is in the correct orientation. If the spacer does not slide in easily or deep, then it needs to be flipped over.

Staying on Track: Making Cogs from Plywood and Dowels

The cogs for this project do not have to be 3D printed. You can utilize the full-scale templates (located in the back of this book or on the website at gitlab.com/MakerMedia/mechengformakers/tree/master) to create patterns to transfer onto ¼" plywood. Be sure to mark each of the pin centers onto the wood with a punch tool (or a nail gently tapped with a hammer). Drill a hole for each pin, ensuring that the center tip of the ¼" drill bit matches up with the marked pin centers. You will be drilling 8 pin holes per spider cog and 34 pin holes on the ring cog.

Once you drill the pin holes, cut the rings out using a band saw, reciprocating saw, scroll saw, hand saw, hobby saw, or rotary tool (whatever you have available). Sand the edges.

Cut at least 75 pins (approximately ⅝" long each) from the ¼" dowel. Insert a dowel pin into each pinhole, seating them with a gentle tap from a rubber

mallet. If you find that any of your pins are loose, simply add a small dab of wood glue to the pin before inserting it. Once complete, they will look like the cogs shown in the image below.

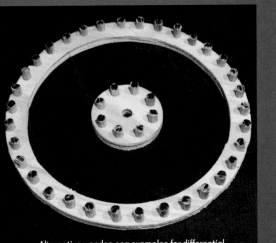

Alternative wooden cog examples for differential

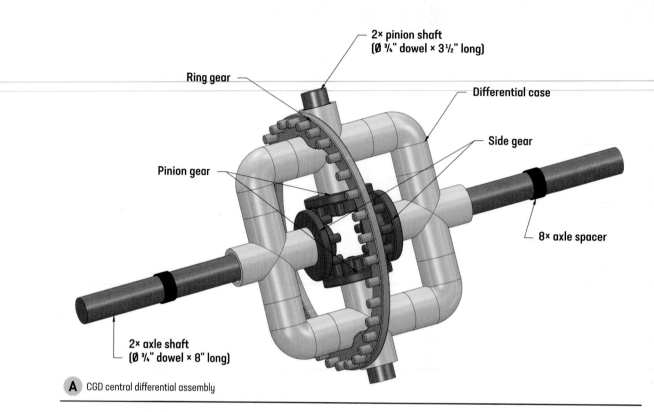

2× pinion shaft
(Ø ¾" dowel × 3½" long)

Ring gear

Pinion gear

Differential case

Side gear

8× axle spacer

2× axle shaft
(Ø ¾" dowel × 8" long)

A CGD central differential assembly

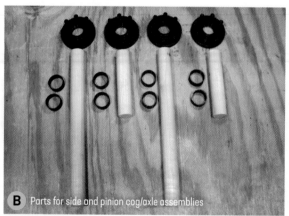

B Parts for side and pinion cog/axle assemblies

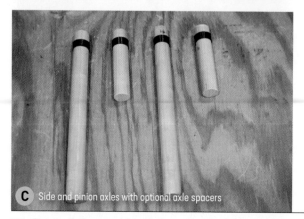

C Side and pinion axles with optional axle spacers

CENTRAL DIFFERENTIAL ASSEMBLY BUILD

The first main module to build is the central differential assembly (Figure **A**). This portion of the differential is comprised of two main side cog (gear)/axle assemblies and two pinion cog (gear)/axle assemblies, all housed in a square frame (differential case), as shown in Figure A.

Begin this part of the build by assembling the two side cog/axle assemblies and the two pinion cog/axle assemblies. All four cogs used in this step have eight teeth. Figure **B** shows all of the parts required to build these assemblies (remember that the spacers are optional).

If you are using the axle spacers, ensure that the spacers are in the proper orientation, and slide one spacer about 1" onto the end of each dowel, as shown in Figure **C**. The second spacer will be installed later after these assemblies are fitted into the differential case.

Now carefully press the axle shafts into the four cogs until the end of each shaft is flush with the inner face of the cog. We find that if you place the cog teeth down on a hard surface, you can tap the axles into place with a rubber mallet from the back. The four final assemblies

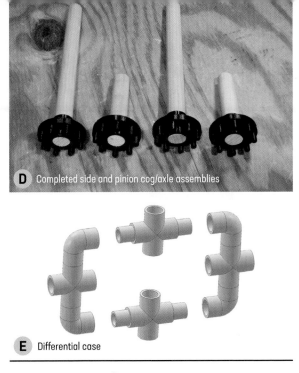

D Completed side and pinion cog/axle assemblies

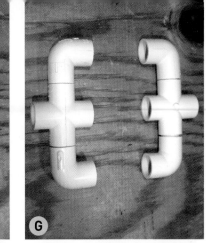

E Differential case

F **G** Differential case side unassembled and left & right sides complete

H Side cog/axle assemblies installed in case sides

I Pinion cog/axle assemblies installed in case top/bottom

J Pinion cog assemblies mating to case side assembly

are shown in Figure **D**.

Now we build something to contain and support the cog/axle assemblies that we just completed. It's time to build the differential case. Figure **E** shows the case in a partially exploded view to guide the build. The case is made up of four ½" schedule 40 elbows (90º), four cross fittings and eight lengths of ½" PVC pipe 1⅜" long each.

Build the left and right sides of the case as shown. These two assemblies are identical, just mirror images of each other (Figure **F** and **G**).

Insert your side cog/axle assemblies into each of differential case sides as shown in Figure **H**. If you are using the axle spacers, ensure that they are properly oriented before sliding the axle assemblies into place.

Now repeat the previous procedure with the pinion cog/axle assemblies with the two remaining cross fittings as shown in Figure **I**. These two assemblies form the top and bottom of the differential case. Note that the two 1⅜" lengths of PVC pipe have been seated in both cross fittings perpendicular to the axle axis.

Now carefully slide the top and bottom assemblies onto one of the side assemblies (Figure **J**). Before fully seating the fittings, make sure that the cog teeth of the pinion cogs are properly engaged with the side cog. Then,

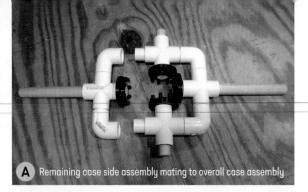

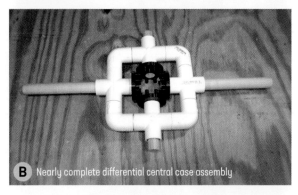

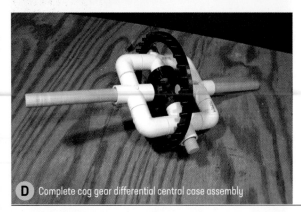

fit the other side of the differential case to the assembly as shown in Figure **A**. Once complete, it will look like Figure **B**.

The final step to complete the differential case assembly is to mount the large ring cog (ring gear). Slide the ring gear onto the differential case as shown in Figure **C**. The gear must be aligned both radially and perpendicularly to the differential case. In the 3D printable version of this gear there are features to help you align the gear radially. However, be careful that the gear is perpendicular to the differential case assembly prior to gluing in place. We have found that ordinary hot glue works quite well to secure the ring gear to the differential case. The complete Cog Gear Differential central case assembly is shown in (Figure **D**).

DIFFERENTIAL FRAME BUILD

Now, let's build the main differential support frame (Figure **E**). We're not going into detail here, since you now have plenty of experience assembling things like this. Build the main frame in three main pieces, as shown in Figure **F**. These three sub-assemblies come together when we assemble the complete differential.

INPUT PINION AND SHAFT CONSTRUCTION

The third and final module is the input pinion cog and axle shaft assembly. Use the same procedures as those used to construct the cog/axle assemblies created for the central differential module. The parts for the input pinion cog and axle assembly are shown in Figure **G**. The axle is made out of ¾" diameter dowel and is 4½" long. Notice some additional parts on the right of the figure. The black, 3D printed part in the upper right of the figure is a ⅜" drill drive adaptor that enables you to drive the CGD with a hand drill. This part is optional, but we did find it quite useful for testing and demonstrating the CGD. The #8 nyloc nut and #8 × 1⅛" socket-head cap screw shown in the picture are for attaching the adaptor. Figure **H** shows the input pinion cog pressed onto its axle with one of the axle spacers slid into place.

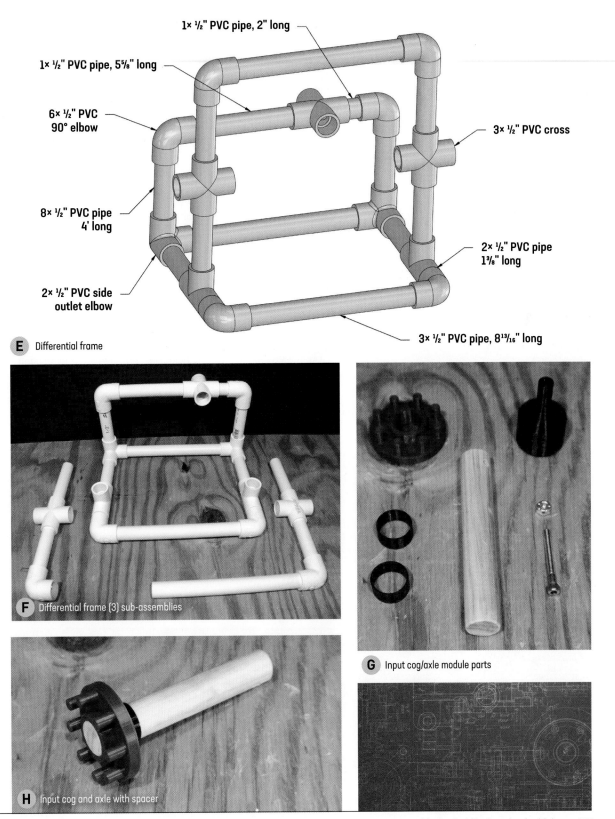

1× ½" PVC pipe, 2" long

1× ½" PVC pipe, 5⅝" long

6× ½" PVC 90° elbow

8× ½" PVC pipe 4' long

2× ½" PVC side outlet elbow

3× ½" PVC cross

2× ½" PVC pipe 1⅜" long

3× ½" PVC pipe, 8¹³⁄₁₆" long

E Differential frame

F Differential frame (3) sub-assemblies

G Input cog/axle module parts

H Input cog and axle with spacer

COG GEAR DIFFERENTIAL FINAL ASSEMBLY

We are now at the point where we can assemble all of our modules into the final, working differential assembly. To start, insert the input cog/axle module into the frame module as shown in Figure **A**. If you are using axle spacers, install a spacer on the input shaft from the outside of the assembly.

If you choose to use the optional drill adaptor, you need to install it now. Supporting the opposite end of the input pinion shaft, push the adaptor all the way until it stops. Next, cross-drill through the adaptor and axle dowel using a ³⁄₁₆" diameter drill bit. Use the side holes in the adaptor as a guide to drill the hole. Secure the adaptor in place using the #8 ×1.5" screw and nyloc nut (Figure **B**).

Slide the two vertical sub-assemblies of the frame module onto the main axles of the central differential module, as shown in Figure **C**. These two vertical parts of the frame module will act as the supports/bearings for the central differential module. Fully seat the horizontal PVC pipe of the vertical sub-assemblies, thereby making a single assembly (Figure **D**).

Finally, install the central differential and support frame assembly into the bottom part of the base frame, being careful to properly engage the ring cog and input pinion cog. Figure **E** shows the completed CGD.

As shown in Figure **F**, you can use a hand drill to test and demonstrate the CGD. Try spinning the input shaft while not touching the output axles. What do you observe? Both axles should spin in the same direction and at the same rate. Now try grabbing one of the output axles so that it cannot rotate. Now, spin the input shaft again. What happens now? The other axle that can still spin should be rotating in the same direction as before, but at a faster rate than it did when both axles were allowed to rotate. Try to figure out how this rotational speed increase is happening. (Hint: gear ratios.)

While we have devoted only one chapter to gears, this topic can easily fill an entire book (or books) with examples and concepts, including the effects materials have on gears and how they are fabricated and used. Gears have been around a long time; primitive versions were made just like the cog gears of our differential

A Input cog/axle module inserted into frame module

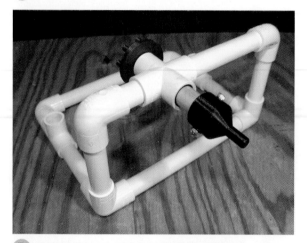

B Input cog/axle with drill adaptor installed

C Vertical frame subs slid onto central differential module

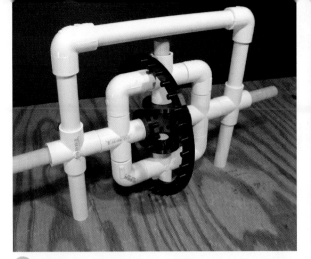

D Vertical frame subs slid with central differential in place

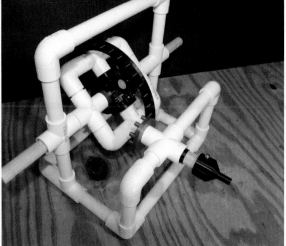

E Complete cog gear differential

F Cog gear differential driven with hand drill

G Wooden Gristmill Gears – notice the use of dowels instead of teeth

example. You can still see them in operation in some old mills, waterwheels, and windmills (Figure **G**).

The content of this chapter should help you understand how to calculate gear ratios, and select the the best types of gears to use in your future projects. We hope that this chapter has served to better acquaint you with this wonderful, powerful, and useful simple machine, the gear.

Staying on Track: Gear Selection

When you have a need for a gear train, keep in mind that it is often easier to repurpose gears that are already available in something else. If you're using gears from different sources, they need to have the same tooth pitch and pressure angle to mesh together. But this can be hard to verify and requires careful measurement if the gears are small. Using a common source for your gears is always your best bet for success.

9 Why Stop Now?

The goal of this chapter is to introduce you to some additional engineering concepts and ideas that can propel your project thinking further, beyond what we have presented in detail in earlier chapters. In essence, we hope that the following discussion makes you ask, "Why stop now?" We are presenting these more advanced concepts via our own personal projects. The first section in this chapter illustrates the final hands-on project in this book, while the other sections are strictly discussions on some of our own past projects. In our project discussions, we have intentionally left out the details of how to build these projects, as they are intended to whet your appetite rather than instruct you on how to make the projects. While we specifically talk about pneumatics, electronics, servos, solenoid sprinkler valves, and microprocessors, the possibilities for expanding your projects are virtually endless. Our goal in sharing these projects with you is to inspire you to experiment on your own, and lead you into new areas of learning.

The Reality of Making vs. Engineering

In this book, we present some great projects, both as teaching tools and to inspire you. We present them in a simple and logical way. But don't get the wrong idea! The reality of these projects, and making in general, is far more chaotic. We've said it elsewhere, but it bears repeating: At least some of the time, you will start down a path to making something, only to find out that your concept is flawed, or you need to build a different way, or there are unintended consequences. This doesn't mean it was wasted time. It just teaches you how not to do it next time. The point is to learn what went wrong and why, and then try again with the new information.

In the engineering world, every attempt is made to anticipate and eliminate failures ahead of time. It takes years to gain the knowledge and skills to do this; and even then, if working with a cutting-edge technology, or pushing the boundaries of human knowledge, the unexpected can happen. But most of the time, engineers spend lots of time analyzing what they make, so that the first one works right away. They may perform a number of experiments, or build models to test ideas before the prototype is made. This is usually because the cost of failure is too high to put a machine, or person, at risk.

On the opposite end of the spectrum, a beginning maker might have to try a dozen times to find success. There's certainly nothing wrong with that, provided one has the time and patience for it. But that's not always practical (or fun). What we hope to do with this book is to give you some fundamental knowledge about engineering. Then you can anticipate some of the more obvious problems you may encounter, and hopefully avoid them by thinking through the execution of your project. Having an understanding of how things can be built, materials, stiffness, strength, modularity, how forces behave, and how energy can be used, are fairly simple concepts that anyone can use to guide their designs toward success. Furthermore, when your design do fail, you'll be better equipped to understand why, and make better versions next time.

A Pneumatic "rockets"

Pneumatic Paper "Rockets"

Pneumatic "rockets" are rockets that are powered by compressed air. They are very inexpensive to make, fun to build, and even more fun to fly - some can travel higher than 300 feet up! Plus, these rockets can be launched over and over again! The body of a pneumatic paper rocket is a tube constructed by rolling a piece of paper around a pipe. The diameter of the rocket body needs to match the diameter of the launch tube that will ultimately be used to fire the rocket. The nose cone for such a rocket can be made out of pretty much anything lightweight: balsa wood, foam, rubber, or paper. Lastly, a flyable pneumatic rocket requires tail fins, which can be made out of nearly any stiff sheet material such as foam core board, cardboard, balsa wood, or thick paper (like cardstock or manilla file folders). Figure **A** shows examples of three pneumatic paper rockets. The orange one on the right has been launched over 100 times!

Technically, by strict definition, these are not rockets. A rocket is generally a cylindrical object that is propelled skyward by some sort of engine and fuel supply contained within itself. The pneumatic rocket that we are talking about here does not have an engine or fuel supply. All of the energy required to launch these rockets skyward comes from air pressure supplied via a launcher. So, technically, our pneumatic rockets are just projectiles that rely on energy from an external source. Having said all of this, we will continue to call them rockets from here forward, as the word rocket just sounds so much cooler than projectile.

So, for our last hands-on project in this book, we show you how to 1) make pneumatic paper rockets and 2) make a pneumatic rocket launcher. We know that this project is not as unique as the others presented in the previous chapters. Just look on the web for pneumatic rocket launchers and you will see many examples. But we still include it as a way to introduce you to pneumatics in a practical, fun way. As you go through this project, think about how your future projects can utilize the power of air pressure. Additionally, we illustrate the use of an electronic **solenoid sprinkler valve** as a way to control fluid (in this case, air) flow; and a **Schrader valve**, which is a one-way check valve used as a "fill" valve for air pressure vessels, such as tires and pneumatic rocket launchers.

Staying on Track: Solenoid Sprinkler Valves

A solenoid sprinkler valve is designed to be used in a lawn sprinkler system, to turn the water flow on and off. This type of valve remains closed, not allowing any water to flow through it when there is no power supplied to it. Only when power is supplied to the valve does it open, allowing water to flow through it.

Now let's talk about a solenoid sprinkler valve in a maker sense. Throughout the book, we discuss repurposing things for your projects. The use of a solenoid sprinkler valve in the pneumatic rocket launcher and pneumatic cannon (discussed later in this chapter) is an excellent example of repurposing. Although originally designed for turning water flow on and off, a solenoid sprinkler valve can also be used to control the flow of air. In its non-energized state, the valve remains closed, keeping the pressurized air within a pressurized vessel to which it is attached. When power is supplied to the valve, it opens, allowing pressurized air to escape out of the pressure vessel.

Solenoid sprinkler valve

Materials:

- » **PVC Pipe – ½" x 18", 1 length**
- » **Regular Paper**
- » **Thick Paper, Card Stock, or Manila File Folder**
- » **Foamcore Board, Cardboard, Coroplast (corrugated plastic sheet)** (or another thin, rigid, lightweight material)
- » **Scissors, or Hobby Knife with a Cutting Surface**
- » **Packing, Electrical, or Aluminum Tape, 1 roll**
- » **Hot Glue Gun and Glue Sticks**
- » **Pencil, Pen or Marker**

MAKING THE ROCKET BUILD PIPE

A rocket build pipe is a way to support and hold your rocket during the building process. It should be the same diameter as the rocket launcher launch tube, since the rocket diameter needs to be snug with the rocket launcher tube. To make your rocket build pipe, cut a piece of pipe the same diameter and length as the launch tube of your rocket launcher. For this project, we used an 18" long piece of ½" PVC pipe. Wrap a single layer of aluminum tape (the type used with ductwork) or a couple of layers of packing tape around the pipe, starting at one end and covering about 14" (Figure **B**). The tape adds a small amount of thickness to the pipe. This added thickness ensures that the rocket body tube has a slightly larger inside diameter than the outside of the rocket launcher launch tube, allowing the rocket to fit snugly against the launch tube, yet slide freely on and off.

B Rocket build pipe

C Rocket body reinforced with packing tape

CONSTRUCTING THE ROCKET BODY

Once the rocket build pipe is complete, wrap a piece of thick paper or 12"×4" cardstock tightly around the build pipe, and wrap the paper with tape over the entire length of the paper (Figure **C**), constructing a tape-reinforced paper tube. Next, "cap" the upper end of the paper tube so that it is airtight and can withstand a sudden blast of pressurized air. We used a piece of cardstock cut in a circle to match the diameter of the tube, with rectangular tabs, as shown in Figure **D**, which extend down the body of the rocket. We secured the cardstock to the end of the tube by taping the tabs down to the tube. Then, we

D Rocket body seal end-cap

wrapped tape around and across the end of the tube using lots more tape, as shown in Figure **A**. The result should be a rocket body completely capped on one end (Figure **B**).

MAKING AND ATTACHING THE ROCKET NOSE CONE

Nose cones are an optional accessory on these paper pneumatic rockets. They make the rockets more able to withstand impacts with the ground, and they make them more aerodynamic (i.e., they have less drag), helping them to fly faster and higher. Construct the nose cone from heavy paper, foam, tape; whatever materials that you have available. The nose cone can be tall and pointed, rounded, bullet-shaped or flat. You can see a few of our constructed nose cones in Figure **C**. Attach the nose cone to the capped end of the rocket body tube with more tape, as shown in Figure **D**.

MAKING AND ATTACHING THE ROCKET FINS

To make the rocket fins, place the rocket body on a piece of paper and draw the outline of the fin shape you want to use on that paper (Figure **E**). Cut out that fin template, and place it on the material you're going to use for your fins. Using the fin template, trace the outline of your fin design onto the fin material. Then, cut 3 or 4 fins from the material. In our case, we cut four fins out of foam core board (Figure **F**).

Installing the fins can be tricky, so we hot glued cardstock "flaps" to the sides of each fin (Figure **G**), and hot-glued the fin spines and flaps to the rocket body, evenly spaced from each other around the circumference of the body. You can further secure the fins with tape, to ensure that they do not fall off during flight (Figure **H**). The vertical placement of the bottom of the fins relative to the body tube should be at the very bottom of the body or slightly below. That way, when the rocket is not on your launch tube, it can stand up on its own on the fins. Also, your rocket will be more aerodynamically stable with the fins mounted as low as possible.

Be aware that if you build your rocket as we have described here, it will be stable and fly just fine; the design inherently takes into account rocket geometry and flight physics. We are not going into details here on things like center of pressure and center of gravity. There is a plethora of information available on the internet about the aerodynamic forces that act upon rockets, and how those forces play into rocket design. We encourage you to do

A Sealing body end

B Sealed body end

C Rocket nose cones

D Rocket nose cone attached to the rocket body

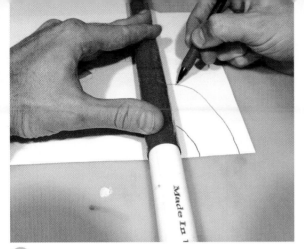

E Rocket fin template

F Rocket fin

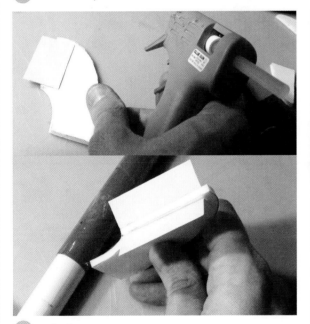

G Rocket fin support

H Rocket with fins installed

some research on your own, and experiment with different nose cone and fin materials and shapes, both practical and bizarre! Also, play with the rocket's center of gravity by adding weight to the nose of the rocket. Some work better than others, while others just look really cool!

DECORATING THE FINISHED ROCKET

Once the fins are installed, take a few minutes to decorate and personalize the rocket. Use markers, tape, stickers, etc. Work with friends and family to make a few rockets, and decorate them individually to add to the fun and competition of launching them (Figure **I**). Now, let's move onto making the launcher! ✪

I Finished rockets with launcher

PNEUMATIC ROCKET LAUNCHER

Materials:

» **PVC Pipe – ½" × 5' – standard schedule 40**
(see cut list for details), **1 length**
(MMC part # 48925K91)

» **PVC Pipe – 2" × 5' – standard schedule 40**
(see cut list for details), **1 length**
(MMC part # 48925K96)

» **PVC Fitting – Elbow (90°) – ½"**, **qty 3**
(MMC part # 4880K21)

» **PVC Fitting – Cross – ½"**, **qty 1**
(MMC part # 4880K241)

» **PVC Fitting – End Cap – 2"**, **qty 1**
(MMC part # 4880K51)

» **PVC Fitting – Coupling – 2" × 2 ⅞"**, **qty 1**
(MMC part # 4880K76)

» **PVC Fitting – Reducer Bushing – 2" × ½"**, **qty 1**
(MMC part # 4880K172)

» **PVC Fitting – Slip Plug – ½"**, **qty 2**
(MMC part # 4880K842)

» **PVC Fitting – Reducer Adapter – ½" × ¾"**, **qty 2**
(MMC part # 4880K433)

» **Solenoid Sprinkler Valve - ⅛" NPT Brass**, **qty 1**
(Lowe's part # 289083)

» **Schrader Valve - ¾"**, **qty 1**
(MMC part # 8063K33)

» **PVC Primer and PVC Cement**, **qty 1 each**

» **Teflon Tape or Other Thread Sealant**, **qty 1 roll**

» **9V Batteries**, **qty 2**

» **Wire**, (such as speaker wire), **~4'**

» **Electrical Tape**, **qty 1 roll**

» **Air Mattress Pump**, **qty 1**

Tools:

» **Saw** (either handsaw, miter saw, or PVC pipe cutting tool)

» **Drill**

» **⅛" NPT Tap or ⅛" Drill Bit**

» **Rubber Mallet**

» **Sandpaper or Sander**

* MMC = McMaster-Carr www.mcmaster.com

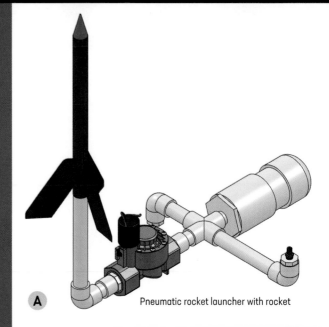

(A) Pneumatic rocket launcher with rocket

In order to propel the paper pneumatic rockets skyward, they must be launched with a sudden blast of compressed air. For the second part of this hands-on project, we are constructing a pneumatic rocket launcher (Figure **A**) - a device that stores a volume of compressed air and then releases the entire volume very quickly into the rocket, sending it up into the wild blue yonder.

CONSTRUCTING THE PNEUMATIC ROCKET LAUNCHER
As with the previous projects in this book, begin the fabrication process of the pneumatic rocket launcher by cutting the PVC pipe lengths per the details in the

CUT LIST

Qty	Material	Cut Length
1	2" PVC pipe	4"
3	½" PVC pipe	2"
1	½" PVC pipe	3"
1	½" PVC pipe	6"
1	½" PVC pipe	18"

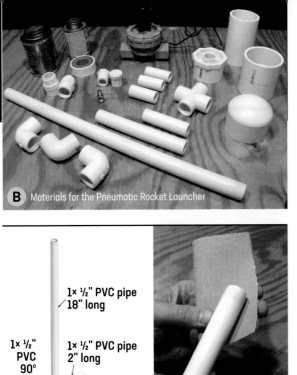

B Materials for the Pneumatic Rocket Launcher

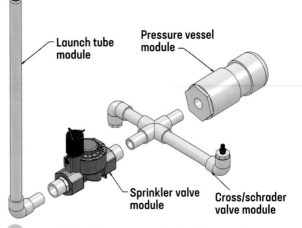

Launch tube module

Pressure vessel module

Sprinkler valve module

Cross/schrader valve module

C Four modules of the pneumatic rocket launcher

1× ½" PVC pipe 18" long

1× ½" PVC 90° elbow

1× ½" PVC pipe 2" long

D Launch tube module

E Using sandpaper to bevel the launch tube

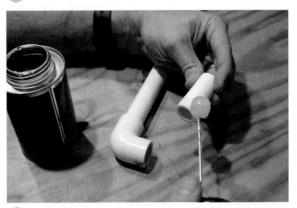

F Using PVC primer/cement to weld PVC

materials/cut list. (If you need a refresher on the best methods of cutting PVC, refer to the *Tracking Further - PVC Pipe Cutting Techniques* on page 34.) The materials and cut pieces are shown in Figure **B**.

Figure **C** shows the four modules of the Pneumatic Rocket Launcher. Each is built separately and then combined to complete the full system.

First, use Figure **D** to collect the parts for the launch tube module. Use sandpaper or an electric sander to bevel or taper one end of the 18" PVC section that will become the launch tube, as shown in Figure **E**. Then, use PVC primer and PVC cement to "weld" the pieces together (Figure **F**), ensuring that the beveled edge of the launch tube is at the top of the assembly. We used a primer/cement combination product for the actual build of this project, but typically the primer and cement are applied separately. Be sure to follow the instructions carefully, use in a well-ventilated area, and work quickly! This stuff dries fast, and once welded together, pieces will NOT come apart. The final launch tube module is shown in Figure **G**.

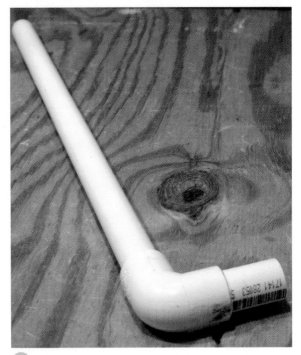

G Launch tube module completed

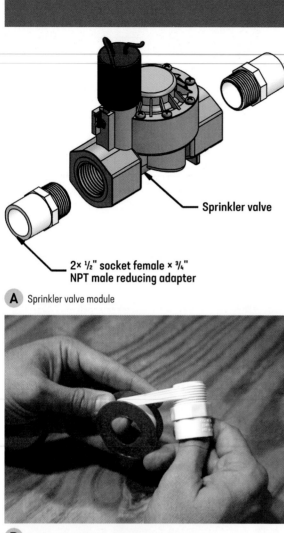

Sprinkler valve

2× ½" socket female × ¾"
NPT male reducing adapter

A Sprinkler valve module

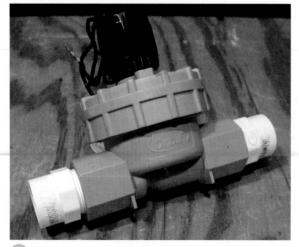

B Applying thread sealant tape

C Sprinkler valve module completed

Next, gather the parts for the sprinkler valve module, as shown in Figure **A**. Wrap the threaded side of each of the reducing adapters with Teflon tape, or another thread sealant (Figure **B**) before tightly screwing them into each side of the sprinkler valve. The completed sprinkler valve module is shown in Figure **C**.

For the cros (Schrader) valve module (Figure **D**), you need to drill and tap one of the slip plugs to accept a Schraeder valve. A Schrader valve is a mechanical, one-way check valve commonly used to fill/hold air in items such as tires on a bicycle or automobile. (If you need details on how to drill and tap PVC pipe, refer to *Tracking Further – Using Ingenuity – Self-Tapping PVC with a Metal Pipe Fitting* on page 40.) Figure **E** shows the tapped plug and valve ready to be installed. When installing the valve, use Teflon tape or another thread sealant on the threads going into the PVC.

As in the launch tube module, all of the PVC pieces in this module need to be welded together with PVC primer and PVC cement, per the configuration shown in Figure D. The completed cross/Schrader valve module is shown in Figure **F**.

The last module is the pressure vessel module, as shown in Figure **G**. These PVC pieces also need to be welded together with PVC primer and PVC cement, per the diagram. The completed pressure vessel module is shown in Figure **H**.

Now that the modules are built (Figure **I**), we need to combine them to create the full launch system. First, connect the pressure vessel module to the cross/Schrader valve module, welding the joint that connects the two modules together with PVC primer and PVC cement. Next, on the sprinkler valve module, look closely at the sprinkler valve and ensure that the arrows printed on the sprinkler unit are pointing away from the pressure vessel and toward the launch tube. These arrows indicate the direction of flow, so we don't want to install it in reverse (although, if that does happen after the joints are welded, you can unscrew the sprinkler unit from the adapters, flip it, and screw it in again). Use PVC primer and PVC cement to weld the joint that connects the cross/Schrader valve module to the sprinkler valve module. Lastly, add the launch tube module to the free end of the sprinkler valve

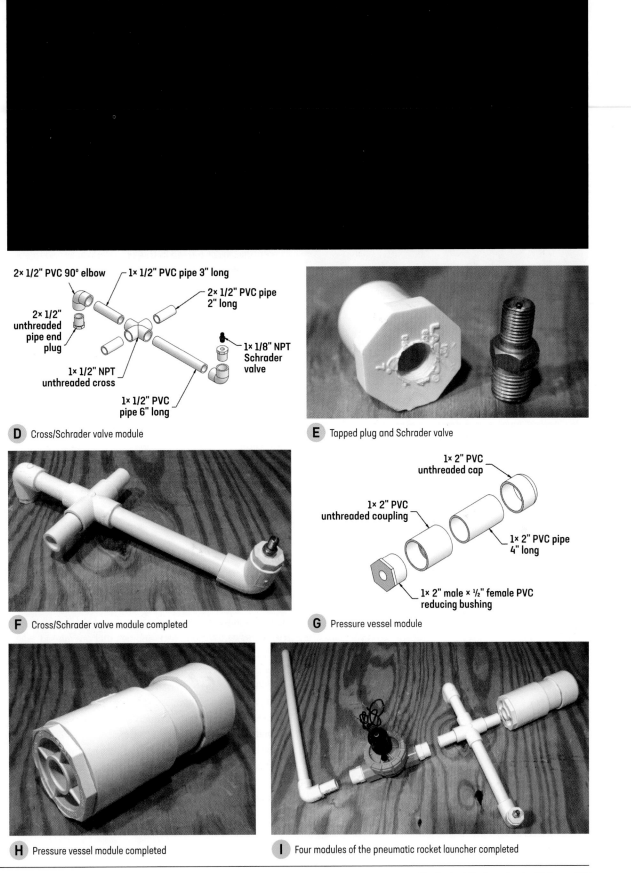

D Cross/Schrader valve module

2× 1/2" PVC 90° elbow — 1× 1/2" PVC pipe 3" long

2× 1/2" unthreaded pipe end plug

2× 1/2" PVC pipe 2" long

1× 1/2" NPT unthreaded cross

1× 1/8" NPT Schrader valve

1× 1/2" PVC pipe 6" long

E Tapped plug and Schrader valve

F Cross/Schrader valve module completed

G Pressure vessel module

1× 2" PVC unthreaded cap

1× 2" PVC unthreaded coupling

1× 2" PVC pipe 4" long

1× 2" male × ½" female PVC reducing bushing

H Pressure vessel module completed

I Four modules of the pneumatic rocket launcher completed

A Pneumatic rocket launcher completed with rocket

B Powering the pneumatic rocket launcher

C Hand-held trigger (optional)

module. Be sure to press the launch tube module into the sprinkler valve module tightly, but DO NOT weld the joint together. We have found that leaving this joint free to move allows the launch tube to be directed at different angles, and removed for easier transport and storage. The combined assembly should look like Figure **A**.

OPERATING THE PNEUMATIC ROCKET LAUNCHER
Before you take your launcher out to a field to test-fire your rocket, you need a way to actuate the pneumatic rocket launcher's (electronic) solenoid sprinkler valve. The sprinkler valve used in this project is designed to open when 24 volts DC (direct current) is supplied to the valve. However, we find that 18 volts is sufficient and simpler to achieve. To get 18 volts, wire two 9-volt batteries together in series. A series circuit is made by wiring the positive terminal of one battery to the negative terminal of another battery. Mechanically, 9-volt batteries make it very easy to "wire" them in series. Simply snap the positive male terminal of one 9-volt battery into the negative female terminal of the second 9-volt battery. Now, wire one of the leads from the sprinkler valve to one of the exposed terminals of the 18-volt battery setup. To actuate the valve, simply touch the other wire from the valve to the second exposed battery terminal. When you do this, you should hear the valve "click," indicating that it did actuate (i.e., open). This setup is shown in Figure **B**.

To take this actuation mechanism a bit further, you can create a hand-held trigger that contains the batteries and wires within it, and fires the solenoid via a pushbutton switch. An example of this system is seen in Figure **C**.

The National Association of Rocketry (NAR, www.nar.org) publishes guidance on model rocket safety. While NAR sanctions rockets with engines that are ignited (electrically or otherwise), some of the guidance offered is useful in this context. This launcher produces an air blast that can cause harm if misused (just like any other machine) at close distances. Always use common sense and never, ever point the launch end at anyone or anything when pressing the trigger. The trigger is the safety valve as well as the launch actuator for this system, and should be accorded the same care you would accord to an air gun.

Figure **D** shows the rocket launcher with a rocket in place, ready for launch. To use your new rocket launcher, start by finding a nice, open area to place your launcher.

Attach a bicycle air pump equipped with a pressure gauge to the Schrader valve, and pump it up to your launch pressure. We recommend that you start low (~20 psi) and move up in small increments, but no more than 80 psi (too much pressure may blow the end cap/nose cone out of your rocket!). Once it is pumped up, stand back and touch the wire to the batteries to send 18 volts to the solenoid sprinkler valve. This discharges the pressurized air from the pressure vessel into the launch tube VERY quickly, and propels the rocket up, up, and away! Based on how much you pressurize the rocket launcher, these little rockets can fly quite far. We have launched them as high as 300 feet or more!

The remainder of this chapter introduces you to some of our own projects. As we state at the beginning of this chapter, the intent of the following discussions is not to instruct you on how to build these projects, but to inspire you to expand your own project knowledge.

Removable/Replaceable Model Rocket Booster Section

Most model rockets implement a recovery system in which the nose cone or the top section of the rocket is separated from the lower section (the booster) via an ejection charge. The ejection charge is ignited at the end of the rocket engine's thrust burn. Sometimes, however, you might want your rocket to remain intact — because you have a large payload, or the rocket itself is meant to continue flying — and dislodging a part of it might expose the payload to forces or temperatures that could damage it. Or, as is the case in this project, you want to just do something different.

Model Rocket Anatomy

- **Body Tubes:** Most smaller model rockets are typically built out of standard-sized tubes made of specially-rolled paper of varying weights. Some, more advanced rocket tubes are made of fiberglass, PVC, or carbon fiber. These body tubes use a numbering system to define them. For example, most of this project uses BT-80 body tubes, which have an inner diameter of about 2.6" and a tube wall thickness of 0.04". This size is typical of "mid-range" rockets with larger or clustered motors. BT-60 body tubes have an inner diameter of 1.6" and a wall thickness of 0.04". The Apogee Rockets site provides details on all the tubing they carry, for example. The benefit of this system is that you can have standard-sized tubing and match a standard fitting nose cone without worrying about fit and finish.

- **Thrust Rings:** A thrust ring is a ring mounted inside a rocket body tube against which the rocket motor pushes to propel the rocket. The thrust ring has to be very securely fixed to the body tube interior and sometimes is integral to the motor mount for the rocket. The top of the motor pushes against this ring, which carries the motor mount and everything attached to it (i.e. the rest of the rocket) through the air.

- **Stuffer Tube:** In a model-rocket design, it may be advantageous to let the hot ejection gases pass through a tube that is inside the main body tube. This is what a stuffer tube is. This allows for the use of the space inside the body tube, between its interior wall and the stuffer tube for electronics, recovery systems, and retractable (such as helicopter recovery rotors) and other mechanical components. Another use of the stuffer tube is to provide an anchor surface for fins that go through the wall of the main body tube — a technique used in high-powered rockets to reinforce the installation of fins. In this project, the fins go through an outer tube and are epoxied to an interior stuffer tube.

Much of this design first appeared in a newsletter published by Apogee Rockets on their website (www.apogeerockets.com) and is referenced here with permission. The newsletter is number 444 in the Apogee newsletter section on their website.

The Apogee design considers a rear-ejection system, and explores the concept of the rocket as a platform. Rear-ejection systems for model rockets present a different set of problems, and are more commonly seen in boost gliders where the rocket can continue flying while the booster stage — or even just the engine casing — returns to earth via a streamer or a parachute. By building a rear-ejection system where the entire fin can is removable, the rocket body can be left as is, but the fin design (shape, size, surface area, etc.) can be varied. The same rocket can then be used for multiple missions (camera platform, telemetry, altitude competition, launch platform, and so on). Adding body-tube sections adds to the rocket's modularity. Different fin cans can also contain single engine or multiple engine mounts, or accommodate larger or different types of engines.

More importantly, if the components are taken to the field for launch after modeling or simulation in software, multiple experiments can be conducted with the same set of components. Arrive at the launch pad and it's super windy (but still legal to fly)? Slide in a fin-can with larger fins to help with stability. Carrying a larger, heavier payload? Grab the three-engine module with the swept forward fins. Better yet, your payloads will never be exposed to hot ejection gases or feel the shock of ejection as the payload bay is hurled forward, away from the booster.

FIN-CAN/CAPSULE

As is the case with most projects, everything starts with a sketch. Here, one design criterion was that the capsule be able to slide in and out of its host body tube easily. It stood to reason that it had to be fabricated primarily out of couplers and a 'stuffer' tube, so that ejection gases do not damage the parachute that is intended to be wrapped around the capsule's middle. Figure **A** shows the initial sketch.

BT-80 tube couplers were used for this just because they happened to be available at the time. If you choose to experiment, feel free to use whatever works for you.

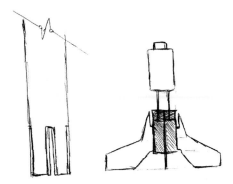

A Rough sketch of the fin capsule and body tube

B Replaceable fin capsule and rocket body tube

The stuffer tube is a BT60 tube left over from another build. The shaded coupler has a pair of centering rings holding the stuffer tube, and the stuffer tube itself has the engine mount in it. There is a single 209mm engine mount in there, which allows for the use of larger motors. In the end, a second coupler as in the depiction was not necessary; a centering ring for the top of the stuffer tube was all that was needed. Ejection gases get funneled up to the stuffer tube to eject the capsule. The fin capsule itself is pushing against the fin slots inside the BT80 tube it is mounted into; The whole lower body tube acts in place of the thrust ring. Just in case, the centering ring at the top of the stuffer tube also pushes against the coupler joining the 2 BT80 slotted body tubes comprising the rocket body. There is enough friction to hold the fin capsule in place on the launch pad (Figure **B**).

C Shock cord mounted to the fin capsule and rocket body tube

The fin capsule slides into the thruster body tube with the fins fitting into pre-cut slots in the tube. For this project, the fins were 3D printed and kept to ⅛" (3mm) thick. The slots themselves provide support for the fins, as well as align the fin capsule inside the main body tube. A cord made out of Kevlar thread or thin rubber cord is attached to both the capsule at the top (to the blue ring as in Figure **C**) and a similar ring inside the main body tube (Figure **D**). The parachute is in turn tied to this cord, and the parachute wrapped around the stuffer tube, before being inserted into the rocket.

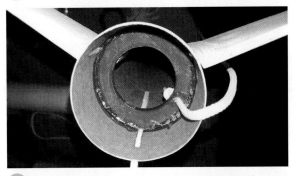

D Shock cord mounted on the ring inside the main body tube

Here are the rocket's components, ready to be put together (Figure **E**). Notice that if desired, the rocket can be made longer by adding more BT80 sections to the top of the rocket. This also allows you to make a length of tube function as an electronics module that you can attach or detach as you see fit.

E Fin capsule and body tube and parachute

Finally, Figure **F** shows the finished rocket, assembled and ready for launch. ❷

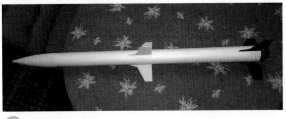

F Assembled rocket

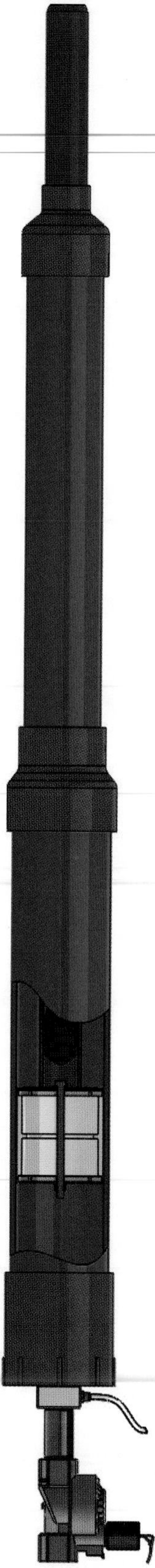

Pneumatic Cannon –
Harnessing the Power of Air

Systems that are powered by compressed or pressurized gasses are known as pneumatic systems, from the Greek *pneuma*, meaning "breath." Although there are many choices of compressed gases, such as nitrogen, carbon dioxide (CO_2), and argon, the typical pressurized gas of choice is air.

In the Pneumatic Cannon, an object (most of the time, a potato) is propelled out of the cannon at a great velocity by pressurized air. Figure **A** shows a CAD model of the Pneumatic Cannon, with a cutaway section so that you can peer inside.

Pneumatic Cannon – Anatomy

PVC plastic pipe and pipe fittings are the primary materials used in the construction of the Pneumatic Cannon. We cover the other materials used in its construction while discussing the anatomy of the cannon in detail.

Figure **B** presents the anatomy of the Pneumatic Cannon, which consists of four primary parts: a barrel, a pressure chamber, a piston, and a solenoid valve. The picture on the left of the figure shows the finished cannon mounted to its adjustable stand. The center image in the figure shows a CAD model of the lower section of the cannon, with a bit of the outside chamber cut away so that you can see inside of it, including the piston. The piston appears in the far right of the figure.

Projectiles to be fired are loaded into (and ultimately launched out of) the barrel of the cannon, which is made of a 4' length of 1½" diameter PVC pipe. The end of the barrel are sharpened by chamfering the inner and outer edge. When a whole potato is jammed onto the sharpened end of the barrel, it cuts out a nice, tight slug of potato that becomes the cannon's projectile. This tightly fitting potato slug can then be pushed down the barrel to its firing position at the barrel's lower end.

The barrel of the cannon is contained within a pressure chamber made up of two larger-diameter PVC pipe sections connected by a PVC pipe fitting. The barrel protrudes a few inches outside of the pressure chamber, and is supported on the upper end of the cannon by a PVC pipe reducer fitting connected to the end of the pressure chamber. The barrel passes through this reducer into

A

3D CAD model of Pneumatic Cannon with cutaway section

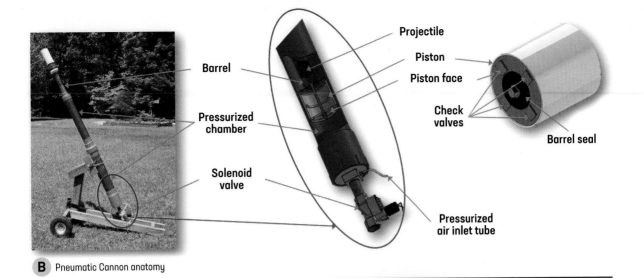

the pressure chamber, and is glued or solvent-welded in place to make an airtight connection. The barrel does not extend all the way into the pressure chamber; there are approximately 12" of free space in the pressure chamber beyond the barrel end. This lower end of the barrel is supported by a ring that resembles the shape of a donut, with extra holes arrayed around it to allow air to pass through it. This support ring was machined out of ¼" acrylic plastic (Figure **C**).

The pressure chamber contains a piston, which can slide back and forth within the pressure chamber in the space beyond the lower end of the barrel. This piston serves as the main component of the primary firing valve for the cannon. The body of the piston is made of a PVC pipe coupler. A disk of acrylic is recessed into each end of the coupler. The disk on the barrel-side of the piston serves as the face of the piston, and the other disk is required solely to provide a means to secure the piston face via a bolt. These disks contain multiple holes to allow air to pass through them. Four small pieces of foam rubber are partially secured to the outside of the piston face, covering the small holes. These foam pieces make simple check valves. Check valves allow something — in this case, air — to flow in only one direction. If air is introduced in the direction of flow of a check valve, it opens the valve and flow through freely. But if the direction of flow is reversed, the check valve closes and stops the flow of air. Also attached to the face of the piston is a large rubber washer, which acts as the barrel seal. The washer makes an airtight seal with the barrel when the piston is pushed into the barrel.

C Pneumatic Cannon barrel support ring

A PVC fitting called a "cleanout" seals the rear or bottom end of the cannon's pressure chamber. A cleanout has a smooth socket on one end that accepts or fits over PVC pipe, while the other end has internal threads. In the case of our Pneumatic Cannon, a pipe plug is screwed into the threaded end of the cleanout, providing a seal for this end of the pressure chamber. When combined with the reducer fitting to seal the barrel to the pressure chamber, this plug makes an airtight pressure vessel. The plug has one small and one large hole drilled into it. The small hole supports an air fitting that supplies the pressurized air to the cannon, and the large hole has a short piece of 1" diameter metal pipe threaded into it. The metal pipe is threaded on both ends with an NPT thread, and these are called pipe nipples.

The last component that makes up the Pneumatic Cannon is a solenoid sprinkler valve, like we used for the Pneumatic Rocket Launcher. This solenoid valve is attached to the cannon by screwing it on to the other end of the pipe nipple protruding out of the pipe plug.

Pneumatic Cannon – How it Works

In simple terms, the Pneumatic Cannon works by exhausting a large amount of air very quickly into a barrel that has been plugged with an object. The air pressure then forces the plug out of the cannon at a very high velocity.

Now we will step through what is going on in the cannon from loading to firing. We will examine the workings of the cannon by looking at the various parts in action, and by analyzing what the pressurized air is doing through those parts at critical points in time.

However, before we get into the nuances of the Pneumatic Cannon we need to talk a bit about air pressure. Air pressure is defined as the force that air exerts on any surface with which it is in contact. In American customary units, air pressure is commonly defined as psi, or pounds per square inch. Uncompressed air is at a particular air pressure known as atmospheric pressure. Atmospheric pressure is defined as the pressure a column of air exerts on a surface due only to the weight of the column of air; it is the air pressure we normally experience all of the time. Typical atmospheric pressure is 14.7psi.

When we talk about compressing air, we generally refer to something known as gauge pressure. Gauge pressure is the air pressure in a container, when measured *greater than* atmospheric pressure. In other words, a gauge pressure of zero is the same as atmospheric pressure. A car tire at 32psi gauge pressure has 32 pounds per square inch *more* pressure than atmospheric pressure.

Armed with the definitions of air pressure, atmospheric pressure, and gauge pressure, we can now look at how the Pneumatic Cannon operates. To understand the operation of the cannon, we must examine three different pressure regions in the cannon along with atmospheric pressure.

The pressure nomenclature is as follows:
- P_{PC} – Gauge pressure of the pressure chamber
- P_B – Gauge pressure of the barrel
- P_I – Gauge pressure of the input pressurized air
- P_{ATM} – Atmospheric pressure

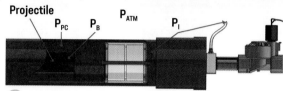

A Pneumatic Cannon operation – stage 1

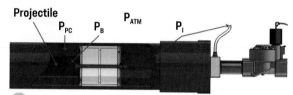

B Pneumatic Cannon operation – stage 2

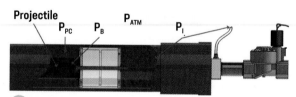

C Pneumatic Cannon operation – stage 3

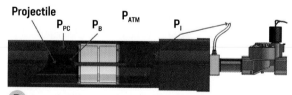

D Pneumatic Cannon operation – stage 4

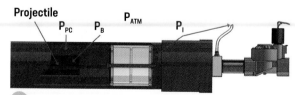

E Pneumatic Cannon operation – stage 5

Pneumatic Cannon Operation - Stage 1

In stage 1 (Figure **A**), the cannon is at rest. The projectile has been loaded and the cannon is at atmospheric pressure in all pressure regions. When at rest, the piston is slid back from the barrel in the pressure chamber. Because the cannon is mounted with some inclination angle, the piston can slide back away from the barrel automatically. The only force the piston sees, in this stage, is that of gravity.

- $P_{PC} = P_B = P_I = P_{ATM}$

Pneumatic Cannon Operation - Stage 2

Stage 2 (Figure **B**) is where we begin the input of pressurized air into the cannon. Pressurized air from an air compressor enters the cannon through a hose attached to the air fitting mounted to the end plug of the pressure chamber. In this stage, only the air pressure P_I on the back or input side of the piston is increased above atmospheric pressure. The increased pressure behind the piston generates a force on the piston driving it up until it is stopped by the back of the barrel. At this stage in the operation of the cannon, the check valves on the face of the piston remain closed, since the pressure on the input side of the piston has not reached a pressure high enough to open the valves.

- $P_I > P_{PC} = P_B = P_{ATM}$

Pneumatic Cannon Operation - Stage 3

As soon as the pressure on the input side of the piston reaches a high-enough value to open the check valves on the face of the piston, we enter Stage 3 (Figure **C**) of the operation. At this point, the pressurized air is able to pass through the piston check valves and into the pressure chamber, increasing the pressure in the pressure chamber P_{PC} that surrounds the barrel. This is what we engineers term a "transient state." A transient state can be defined as a period of time where energy distribution in a system is actively changing values or locations. In the case of our Pneumatic Cannon at Stage 3, the maximum pressure is on the input or back side of the piston. The pressure chamber is also increasing in pressure, though it is still not as high as the input side. However, the pressure chamber pressure P_{PC} is greater than the barrel pressure P_B (which at this stage of the operation is still at atmospheric pressure because the barrel is sealed off from the pressure chamber by the piston).

- $P_I > P_{PC} > P_B = P_{ATM}$

Pneumatic Cannon Operation - Stage 4

Stage 4 (Figure **D**) represents the cannon at a steady state in terms of pressure. The pressure in the pressure chamber P_{PC} is equal to the pressure on the inlet side of the piston P_I. Both the pressure chamber pressure and the inlet side pressure are much higher than the barrel pressure P_B, which is still at atmospheric pressure P_{ATM}. Considering that the pressures on both sides of the piston are equal, there is no force on the check valves on the face of the piston to hold them open. The pressure on each side of the piston is much greater than the barrel, so there is a suction force generated holding the piston against the barrel. This suction force overcomes the weight of the piston, keeping it in place. At this stage, the cannon is ready to fire.

- $P_B = P_{ATM}; P_I = P_{PC} >> P_{ATM}$

Pneumatic Cannon Operation - Stage 5

Stage 5 is the firing stage (Figure **E**). The solenoid sprinkler valve is energized, causing it to open. This allows the pressurized air on the back or inlet side of the piston to be rapidly released. The pressure on the front or pressure chamber side of the piston P_{PC} is now much higher than on the inlet side P_I, generating a force that drives the piston rearward, away from the back of the barrel. The back end of the barrel is now completely open to the pressure chamber. The high-pressure air from the pressure chamber exits through the barrel, which is plugged by our projectile. This sudden flux of high-pressure air impinging on the back of the projectile creates a large force on the projectile. This force accelerates the projectile along the length of the barrel until it exits the cannon at a rather high velocity. The cannon has fired!

- $P_B = P_{PC} > P_I = P_{ATM}$

A Altitude Dropper attached to drone

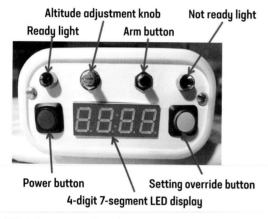

Ready light · Altitude adjustment knob · Arm button · Not ready light

Power button · 4-digit 7-segment LED display · Setting override button

B Altitude Dropper control panel

C Altitude Dropper 4-digit 7-segment LED display board (front)

D Altitude Dropper 4-digit 7-segment LED display board (rear)

Altitude Payload Dropper – Using Hobby Servos, Sensors & Microcontrollers

Some projects can be greatly enhanced by —or even require a bit of — intelligence. A project's intelligence can come in the form of these nifty little low-cost computers known as microcontrollers. Once your project has a microcontroller integrated with it, you can do all sorts of cool things. You can add servos and sensors to have your project move in response to events in its environment, or blink lights and respond to inputs from you, the user. The Altitude Payload Dropper project is an example of a project that integrates a microcontroller with a control panel and servos to make things move. Our intention in designing and building this project was to make a mechanism that could be carried up to a preset height by a quadcopter, and then open a set of bomb bay doors to drop a payload (a model parachute man, fake toy bomb, etc.). The Altitude Payload Dropper had to be small and light enough to strap to the belly of a model aircraft such as a drone (Figure **A**).

Altitude Payload Dropper – Anatomy

Figure **B** shows the control panel of the Altitude Payload Dropper. The control panel is made up of four indicator lights, three switches, an adjustment knob, and a large central display. These components are mounted on a faceplate made from thin plywood.

Central to the control panel is a display. This specific display used for this project is known as a 4-digit, 7-segment LED display. As you can see in Figure **C**, there are four "8s" representing the four digits, and each digit is made up of seven individual segments. The segments are actually LEDs (light emitting diodes) that work like small light bulbs. By selectively powering these LED segments, thus lighting them up, we can generate any numeric character between zero and nine. There are six additional small, circular LED (dots) on the display between the digits. The four dots along the bottom are for illuminating a decimal place, and the two vertical dots in the middle allow the display to be used as a digital clock.

Figure **D** is a view of the back of the 4-digit, 7-segment display. The display is mounted to a circuit board with a large computer chip in the middle; this display and board combination is known as a backpack. The backpack is designed to simplify the interfacing of the display into your project, from a physical and programmatic standpoint.

E Altitude Dropper barometric pressure sensor

F Altitude Dropper bomb bay door servo

Most microcontrollers operate at 5 volts. Most LEDs operate at 2.1 volts. An LED requires a resistor to limit the voltage; without a resistor, the LED would immediately burn out from the excessive voltage. With an un-backpacked display, you would need to wire a resistor to each individual LED segment and dot of the display. That would be 34 resistors! Then, to display a particular number on a 7-segment display, you must power the correct LED segment combination to generate that number. From a programmatic standpoint, this is quite arduous.

With the backpack, all of the required resistors are integrated into the board's computer chip, so you don't have to painstakingly add a resistor to each LED you intend to use. The computer chip on the backpack can receive the number that you need to display directly from your program. It then internally determines which LED segments to power to display the number.

The Altitude Payload Dropper uses a barometric pressure sensor, also known as an atmospheric pressure sensor, to determine a change in altitude (Figure E). As you gain altitude, the barometric pressure decreases at a known rate. By monitoring the change in barometric pressure (as read by the pressure sensor), the Altitude Payload Dropper can calculate how far it has risen above the ground.

Bomb bay doors open once the Altitude Payload Dropper reaches a preset altitude above the ground. The actuation or opening of the doors is achieved by a small hobby servo (Figure F) mounted to each of the two hinged bomb bay doors. A hobby servo is a pretty cool little device. These

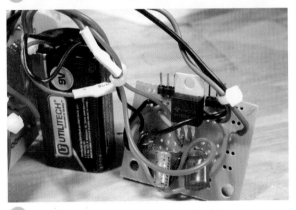

G Altitude Dropper power management board with battery

servos are used in many applications, such as in model airplanes to move the control surfaces, in model cars to steer the wheels, and in robots to move joints and grippers.

A servo is made up of a small electric motor, a gearbox, and a sensor that can precisely monitor rotation. Through simple commands from the microcontroller, a servo can be moved to any particular angle. You can also tell the servo how fast to accelerate and how fast to move to a certain angular position. Once the Altitude Payload Dropper reaches the preset altitude above the ground, it sends a command to the servos attached to the bomb bay doors that then opens them to a predetermined, precise angle.

Servos require a steady, known DC voltage to operate properly. Figure G shows the power management board that was designed specifically for the Altitude Payload Dropper. The input voltage to the Dropper is supplied

A Altitude Dropper arduino microcontroller

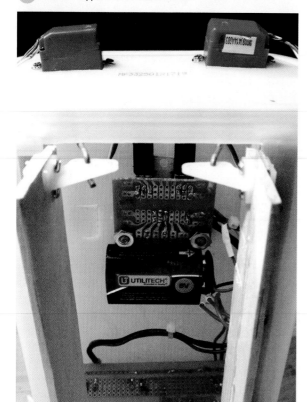

B Altitude Dropper view from bottom, bomb bay doors open

by a 9V battery. The servos used in this project require a voltage of 4.8 volts (this value was rounded to an even 5 volts.) The output voltage of 5 volts is controlled by a small computer chip called a linear voltage regulator, which is the black rectangular device toward the middle of the board shown in the figure. The silver metal tab protruding from the top of the chip is a heatsink. During operation, the chip heats up, and the heatsink helps the chip dissipate some of this heat into the air around it.

The "brains" for this project are in the form of a small computer known as a microcontroller. Specifically, the Altitude Payload Dropper uses an Arduino microcontroller (Figure A). A microcontroller is programmed from a personal computer, using a programming language specific to the microcontroller. All of the control panel inputs, such as button presses and adjustment knob rotation, as well as the display functions, sensor readings, and servo actuations, are controlled by the Arduino "brains" of the Altitude Payload Dropper.

The display, wires, boards, chips, servos, and sensors are housed in a repurposed plastic baby wipe container (Figure B). You can see the two blue servos that are used to open the bomb bay doors at the top of the figure. The bomb bay doors are made from lightweight balsa wood and hinged from a rectangular opening cut into the baby wipe container.

Altitude Payload Dropper – How it Works
When you first power up the Altitude Payload Dropper by pressing the red power button, the bomb bay doors automatically open and the microcontroller goes into "altitude adjust mode" (Figure C). The red "not ready" light will be illuminated in this mode. Altitude adjust mode allows the user to select a specific altitude, in feet, above the ground that the Altitude Payload Dropper will open its doors, thereby dropping whatever is inside. You rotate the altitude adjustment knob until the desired altitude is displayed on the LED display. In figure C the drop altitude is set for 200 feet.

Once the desired drop distance (the altitude above the ground) has been set, it is time to load the payload. To load the payload, the Altitude Payload Dropper has to be turned upside down, so that the payload can be dropped into the payload compartment within the belly of the Dropper. To close the bomb bay doors, you press the arm button on the control panel. Pressing the arm button puts the Altitude Payload Dropper into "ready for flight mode" and illuminates the green "ready light" (Figure **D**).

Not everyone lives at sea level. If you already live at an altitude of, say, 500 feet, you're well above the 200-foot trigger limit you've just set. To remedy this, when the arm button is pressed, the Arduino microcontroller and the pressure sensor record the current baseline ground level reading (378 feet in this example.) While in the ready-for-flight mode, the Altitude Payload Dropper is constantly monitoring the altitude, and comparing this altitude reading to the initial ground level reading that was recorded the instant that the arm button was pressed. The microcontroller takes the most recent altitude reading and then subtracts the ground level reading from it. It does this comparison over and over until the difference is equal to or greater than the drop altitude that was set in the altitude adjust mode. At this point, the microcontroller sends a signal to the servos to open the bomb bay doors, dropping the payload. In our current example, the bomb bay doors will open when the Altitude Payload Dropper is flown 200 feet above the ground.

A nice feature of the Altitude Payload Dropper is the ability to change the initially-set drop height after the payload has been loaded and the Dropper armed, without opening the bomb bay doors. This is where the setting override button comes into play. The drop altitude can be changed by holding this button down, and rotating the altitude adjustment knob until the new altitude desired is indicated on the display. Once the override button is released, the Altitude Payload Dropper is re-armed and returns to the ready-for-flight mode (Figure **E**).

We hope that you enjoy the topics and projects we have detailed in this book, and that they provide you with additional insight and tools to make your projects bigger, better, stronger, faster, and able to go further. You now should have a good grasp on how to transition from an idea to a design, and from there to a prototype and a final project. Along the way, we have introduced you to the idea of modularity, and how to find materials and components from common and perhaps not-so-common sources, as well as given you some ideas on how to repurpose a variety of things to suit your project needs.

You have also seen how to make the tools you need to

C Altitude Dropper powered on – altitude adjust mode

D Altitude Dropper powered on – ready for flight mode

E Altitude Dropper powered on – altitude override mode

accomplish your goals, and learned the utility of simple machines. We have shown you how to select materials for your project, how to keep all the parts together, and how to provide your machines with the means to make them move. We have, in short, tried to bring you closer to being an engineer than you may have been before. No matter what your project is, however large it may be, if you break it down into small enough parts and consider each as a system within a larger system, you can design and build anything — time and cost permitting, of course. Do not allow the complexity or size of your particular idea be a deterrent. If you keep at it, no matter how long it takes, eventually your design will become reality.

HAPPY MAKING!

Index

B

C

D

E